T0212207

Human Computation

Synthesis Lectures on Artificial Intelligence and Machine Learning

Editors
Ronald J. Brachman, *Yahoo Research*
William W. Cohen, *Carnegie Mellon University*
Thomas Dietterich, *Oregon State University*

Human Computation
Edith Law and Luis von Ahn
2011

Trading Agents
Michael P. Wellman
2011

Visual Object Recognition
Kristen Grauman and Bastian Leibe
2011

Learning with Support Vector Machines
Colin Campbell and Yiming Ying
2011

Algorithms for Reinforcement Learning
Csaba Szepesvári
2010

Data Integration: The Relational Logic Approach
Michael Genesereth
2010

Markov Logic: An Interface Layer for Artificial Intelligence
Pedro Domingos and Daniel Lowd
2009

© Springer Nature Switzerland AG 2022

Reprint of original edition © Morgan & Claypool 2011

All rights reserved. No part of this publication may be reproduced, stored in a retrieval system, or transmitted in any form or by any means—electronic, mechanical, photocopy, recording, or any other except for brief quotations in printed reviews, without the prior permission of the publisher.

Human Computation

Edith Law and Luis von Ahn

ISBN: 978-3-031-00427-8   paperback
ISBN: 978-3-031-01555-7   ebook

DOI 10.1007/978-3-031-01555-7

A Publication in the Springer series
SYNTHESIS LECTURES ON ARTIFICIAL INTELLIGENCE AND MACHINE LEARNING

Lecture #13
Series Editors: Ronald J. Brachman, *Yahoo Research*
                William W. Cohen, *Carnegie Mellon University*
                Thomas Dietterich, *Oregon State University*
Series ISSN
Synthesis Lectures on Artificial Intelligence and Machine Learning
Print 1939-4608   Electronic 1939-4616

Human Computation

Edith Law and Luis von Ahn
Carnegie Mellon University

SYNTHESIS LECTURES ON ARTIFICIAL INTELLIGENCE AND MACHINE LEARNING #13

ABSTRACT

Human computation is a new and evolving research area that centers around harnessing human intelligence to solve computational problems that are beyond the scope of existing Artificial Intelligence (AI) algorithms. With the growth of the Web, human computation systems can now leverage the abilities of an unprecedented number of people via the Web to perform complex computation. There are various genres of human computation applications that exist today. Games with a purpose (e.g., the ESP Game) specifically target online gamers who generate useful data (e.g., image tags) while playing an enjoyable game. Crowdsourcing marketplaces (e.g., Amazon Mechanical Turk) are human computation systems that coordinate workers to perform tasks in exchange for monetary rewards. In identity verification tasks, users perform computation in order to gain access to some online content; an example is reCAPTCHA, which leverages millions of users who solve CAPTCHAs every day to correct words in books that optical character recognition (OCR) programs fail to recognize with certainty.

This book is aimed at achieving four goals: (1) defining human computation as a research area; (2) providing a comprehensive review of existing work; (3) drawing connections to a wide variety of disciplines, including AI, Machine Learning, HCI, Mechanism/Market Design and Psychology, and capturing their unique perspectives on the core research questions in human computation; and (4) suggesting promising research directions for the future.

KEYWORDS

human computation, human-in-the-loop algorithms, output aggregation, active learning, latent class models, task routing, labor markets, games with a purpose, task design, crowdsourcing, mechanism design, incentives

To Eric, Jacob and my parents.

– Edith

To Laura.

– Luis

Contents

Preface

Human computation is a new and evolving research area that centers around harnessing human intelligence to solve computational problems (e.g., image classification, language translation, protein folding) that are beyond the scope of existing Artificial Intelligence (AI) algorithms. With the growth of the Web, human computation systems can now leverage the abilities of an unprecedented number of people via the Web to perform complex computation. There are various genres of human computation applications that exist today. Games with a purpose (e.g., the ESP Game) specifically target online gamers who generate useful data (e.g., image tags) while playing an enjoyable game. Crowdsourcing marketplaces (e.g., Amazon Mechanical Turk) are human computation systems that coordinate workers to perform tasks in exchange for monetary rewards. In identity verification tasks, users perform computation in order to gain access to some online content; an example is reCAPTCHA, which leverages millions of users who solve CAPTCHAs every day to correct words in books that optical character recognition (OCR) programs fail to recognize with certainty.

Despite the variety of human computation applications, there exist many common core research issues. How can we design mechanisms for querying human computers such that they are incentivized to generate truthful outputs? What are some techniques for aggregating noisy or complex outputs from multiple human computers in the absence of ground truth? How do we effectively assign tasks to human computers in order to satisfy the objectives of both the system (e.g., quality, budget and time constraints) and the workers (e.g., desire to succeed, to learn, to be entertained)? What classes of computational problems can be efficiently answered using human computation? What are some programming paradigms for designing human computation algorithms? How can human computation systems leverage the joint efforts of both machines and humans?

This book is aimed at achieving four goals: (1) defining human computation as a research area; (2) providing a comprehensive review of existing work; (3) drawing connections to a wide variety of disciplines, including AI, Machine Learning, HCI, Mechanism/Market Design and Psychology, and capturing their unique perspectives on the core research questions in human computation; and (4) suggesting promising research directions in the field. Additional supplementary materials for this book (including the slide for the AAAI 2011 tutorial on "Human Computation: Core Research Questions and State of the Art") can be found at `http://humancomputation.com/book`. We hope that this book will be a useful resource in the years to come for both newcomers and seasoned researchers who are interested in human computation, or more generally, the study of computational systems with humans in the loop.

Edith Law and Luis von Ahn
July 2011

Acknowledgments

As a relatively young field, the idea of human computation is not yet well defined. This book is in part an accumulation of ideas from people in the field with whom we had discussions and debates about the definitions, scope and future directions of human computation. These people include Paul Bennett (Microsoft Research), Eric Blais (Carnegie Mellon University), Yiling Chen (Harvard), David A. Grier (George Washington University), Eric Horvitz (Microsoft Research), Panos Ipeirotis (New York University), Winter Mason (Yahoo! Research), Rob Miller (MIT), Tom Mitchell (Carnegie Mellon University), David Parkes (Harvard University), Paul Resnick (University of Michigan), and Haoqi Zhang (Harvard University). Last, but not least, we thank our families for their unfailing support throughout this time-consuming, but invaluable, project.

The writing of this book is supported by the National Science Foundation Social-Computational Systems (SoCS) program under Grant No. IIIS-0968487. Any opinions, findings, and conclusions or recommendations expressed in this material are those of the authors and do not necessarily reflect the views of the National Science Foundation.

Edith Law and Luis von Ahn
July 2011

CHAPTER 1

Introduction

1.1 COMPUTATION: NOW AND THEN

While *computation* is a central concept in computer science, there is much debate about its exact definition. Originating from the Latin word *computare*, to compute is to "count, sum up or reckon together." This simple definition of computation is not a far removed description of what early *computers* actually did. Inspired by the use of holes punched onto train tickets to record information about passengers, Hollerith invented the tabulating machine, which used punched cards to record data in a way that was readable by machines. The Hollerith Machine was used to compute statistics about population during the 1890 census in the United States, summing up data about more than 75 million individuals, such as the number of people who were married, in each profession, parents to certain number of children, or speakers of English.

However, long before any modern desktop computers or tabulating machines, computation was carried out by *humans* [131]. In fact, before the adoption of tabulating machines, the word *computer* referred to a person who performed calculation as a profession, and many tabulating machines that came about later on were named using acronyms that ended in "AC," meaning "Automatic Computer," [89] in order to be distinguished from *human computers*. Figure 1.1 shows the historic timeline of some computation projects undertaken by human computers, as described in [131].

The so-called *organized computation* projects [131] involved anywhere from a few human computers (e.g., the computation of the trajectory of the Halley's Comet in 1758 was undertaken by only three highly skilled astronomers) to hundreds of individuals (e.g., the Mathematical Table Project led by Arnold Lowan and Gertrude Blanch in 1938 employed more than 450 people unskilled in mathematics), and were created to tackle scientific or practical problems that the society faced at the time, e.g., creating decimal trigonometry tables in order to support the metric system, calculating the trajectory of bombs during the first World War, or modeling the stock market during the depression in 1930.

There are several key concepts in organized computation that we will use to define what is meant by *computation* and, more importantly, *human computation*. First, a complex problem is decomposed into basic operations, distributed to many individuals in an organized way, and re-assembled to reach a solution. For example, to measure the quantity of interest (e.g., the trajectory of a comet), a scientist or mathematician would devise a mathematical formula, break down the formula into a set of simpler quantities that can be easily computed by an individual human computer, then re-assemble the results. Second, computation is carried out using an explicit set of instructions, leaving little to interpretation. Human computers were often given a "computing plan"—a sheet of

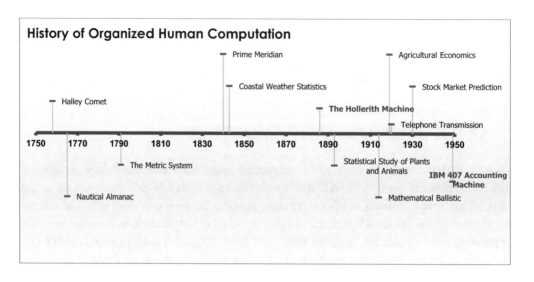

Figure 1.1: Timeline of organized human computation [131].

paper with explicit instructions for each step of the computation—to follow. These computing plans contain *algorithms*, i.e., finite sequences of explicit instructions to transform input to output, as well as procedures for detailed numerical and error analysis. Finally, similar to human computation systems today, accuracy and efficiency (in terms of time and cost) were two important criteria that project leaders strived for. For example, to lower the chances of mistakes, computation was sometimes done by two independent human computers, and checked by a third person who compared the results. For efficiency, mathematical techniques such as interpolation were used to speed up the process of computation.

The story of organized computation is quite different today. Most computational problems that required human processing in the past can now be trivially solved using computers. Mathematical problems that previously would have taken many human computer months of work (e.g., the calculation of correlation coefficients, regression analysis, and differential equations, etc.) can now be done in seconds. However, despite the power of the modern computers, there are still many problems that are easy for humans to solve, but difficult for even the most sophisticated computer algorithms. The problems that fall into this category can be referred to as Artificial Intelligence (AI) problems. Such problems include perceptual tasks (e.g., object recognition, music classification, protein folding), natural language analysis (e.g., sentiment analysis, language translation) and complex cognitive tasks (e.g., planning and reasoning). To date, there are no algorithms that can reach or exceed human level performance on many AI problems. Moreover, many of these algorithms are based on machine learning approaches that require large amounts of training data, which are time-consuming and expensive to obtain. Usually, training datasets are assembled by having a few paid volunteers manually and meticulously annotate the objects of interests (e.g., text, music, images and videos), essentially acting as human computers.

The scale of today's organized computation is no longer restricted to tens to hundreds of human computers, but instead millions of, if we can somehow leverage the Internet and the activities of its growing user base. As a by-product of everyday activities on the Web, users are generating a wealth of information about objects that researchers are interested in analyzing. For example, there is no shortage of users tagging images, music and videos at popular collaborative tagging websites such as Flickr, Last.FM and YouTube, expressing their opinions in blogs and tweets, and contributing their knowledge in Wikipedia and Question and Answer forums. Among these web applications are new *crowdsourcing* systems, such as Mechanical Turk, which enable large-scale, distributed organized computation that involves thousands of workers performing tasks in return for monetary payments.

These new trends and systems present exciting new opportunities for AI: we can now build automated systems that leverage the computation of a huge number of human computers (which we call "human computation systems") to tackle problems that neither humans nor machines alone can solve easily. The goal of this book, and of research in Human Computation, is to shed some light on the anatomy of these human-in-the-loop computational systems, methods for optimizing their efficiency and effectiveness at solving problems, and the fundamental research questions that need to be addressed in building such systems.

1.2   WHAT IS HUMAN COMPUTATION?

Since 2005, the phrase "human computation" [332] has become synonymous with many other equally loosely defined research areas, such as "crowdsourcing," "social computing," "socio-computational systems," and "collective intelligence." In clearly defining what we mean by "human computation," we can outline the scope of this research area, pinpoint the fundamental research questions, identify end goals and focus our efforts in reaching them.

To understand what we mean by "human computation," we must first define the word "computation." In our formulation, **computation** is the process of mapping of some input representation to some output representation using an explicit, finite set of instructions (i.e., an *algorithm*). In the classic work by Alan Turing, computation is similarly defined, where the input and output representations are symbols, the process is the writing of symbols in each cell of an unlimited tape, and the set of instructions or algorithm is a state transition table that determines what symbol should be written for the current cell. Similarly, a human computer who is given two quantities (input representation) and asked to multiply them together (explicit instruction) to generate a product (output representation) is performing computation. In fact, the Turing Machine was meant to mimic the capability of human computers in carrying out mathematical calculations. In Turing's own words, "the idea behind digital computers may be explained by saying that these machines are intended to carry out any operations which could be done by a human computer" [321].

Following this definition, **human computation** is simply computation that is carried out by humans. Likewise, **human computation systems** can be defined as intelligent systems that organize humans to carry out the process of computation—whether it be performing the basic operations (or units of computation), taking charge of the control process itself (e.g., decide what operations to execute next or when to halt the program), or even synthesizing the program itself (e.g., by creating new operations and specifying how they are ordered). The meaning of "basic" varies, depending on the particular context and application. For example, the basic unit of computation in the calculation of a mathematical expression can be simple operations (such as additions, subtractions, multiplications and divisions) or composite operations that consist of several simple operations. On the other hand, for a crowd-driven image labeling system, a user who generated a tag that describes the given image can also be considered to have performed a "basic" unit of computation. In an experiment for solving the graph coloring problem using distributed human computation [172], the basic unit of computation each human solver performed was to "change the color of his or her node, given the current colors of the neighboring nodes." The system's job was to simply take the output from each human solver and update the state (i.e., color of each node) of the network.

Finally, there exist many so called *volunteer computer* projects, running on the BOINC [38] framework, where people donate their idle CPU power to help solve large computational problems. SETI@Home [283] uses home computers to detect signs of extra-terrestrial intelligence by pin-pointing narrow-bandwidth radio signals. PrimeGrid [260] is a BOINC project for finding prime numbers for various mathematical problems. Einstein@Home [100] is a volunteer computing project for detecting gravitational waves from spinning neutron stars (called pulsars), leading to new discoveries [243]. Similar efforts have been directed towards solving physics problems (LHC@Home [197]), folding proteins (Folding@Home [114]), and solving puzzles such as Sudoku. There also exist *participatory sensing* [47] projects, where humans carry mobile devices mounted with sensors to help collect data (e.g., about the environment) at different geographic locations. By our definitions, these are not considered human computation systems because they do not require humans to perform computation themselves. For example, a participatory sensing project where humans actively decide where and when to collect information can be considered human computation; on the other hand, it is not human computation if the participants are merely the sensor carriers, with no **conscious** role in determining the outcome of the computation.

1.2.1   EXPLICIT CONTROL

How is the concept of human computation related to other concepts, such as crowdsourcing, collective intelligence and social computing? In the spirit of crowd wisdom, let's first examine these concepts as they are defined in Wikipedia (Figure 1.1).

Based on these definitions, "crowdsourcing" can be considered a method or a tool that human computation systems can use to distribute tasks through an open call. However, a human computation system does not need to use crowdsourcing; a system that assigns tasks to a closed set of workers

	Table 1.1: Related Concepts [352]
Crowdsourcing	The act of outsourcing tasks, traditionally performed by an employee or contractor, to an undefined, large group of people or community (a crowd) through an open call.
Collective Intelligence	A shared or group intelligence that emerges from the collaboration and competition of many individuals and appears in consensus decision making in bacteria, animals, humans and computer networks.
Social Computing	Technology for supporting any sort of social behavior in or through computational systems, e.g., blogs, email, instant messaging, social network services, wikis and social bookmarking. Technology for supporting computations that are carried out by groups of people, e.g., collaborative filtering, online auctions, prediction markets, reputation systems, computational social choice, tagging and verification games.

hired through the traditional recruitment process (e.g., resumes, in-person interviews) can still be considered a human computation system.

The term "social computing" is a broad concept that covers everything to do with social behavior and computing. Human computation intersects social computing in that some, but not all, human computation systems require social behavior and interaction amongst a group of people. That is, human computation does not necessarily involve large crowds, and workers are not always required to interact with one another, either directly or indirectly (e.g., through a market mechanism).

Finally, "collective intelligence" refers to the emergent intelligent behavior of a group of individuals, which includes non-humans and non-living things. Collective intelligence, therefore, is an even broader concept that subsumes crowdsourcing, social computing and human computation. Consult [91] for a detailed survey and alternative perspectives on Web-based mass collaboration systems.

Most importantly, none of the related concepts emphasize the idea of **explicit control**. In fact, they assume that a large part of the computational outcome is determined by the natural dynamics (e.g., coordination and competition) [177, 179] between the individuals of a group, which the system cannot or does not deliberately control. There is no explicit decomposition or assignment of task, nor are there any explicitly designed mechanisms for ensuring that the human computers tell the truth. In reality, the amount of explicit control in crowd-driven systems is a continuum. For example, many crowd-driven systems (e.g., Wikipedia) do enforce rules, protocols and standards (such as "technical specifications" or "routinized processes" [214]) by which individuals need to abide. What is different about human computation systems is the level of explicit control, which is on the greater end of the spectrum. In other words, instead of focusing on studying human behavior, the focus of human computation research is on *algorithms*, which either specify exactly what gets processed, by

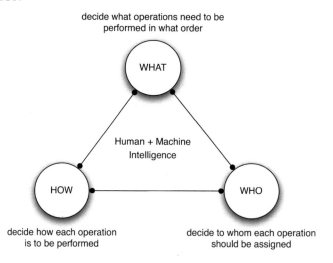

Figure 1.2: Three central aspects of human computation systems.

whom and how, *or* explicitly organize human efforts to solve the problem in a well-defined manner. In Chapter 2, we will examine the idea of *human computation algorithms* in more details.

Conceptually, there are three aspects—"what," "who" and "how"—of any human computation systems (depicted in Figure 1.2) where explicit control can be applied.

The "What" Aspect

In order to generate a solution to a computational problem, we must have an algorithm that outlines exactly how to solve the problem. An algorithm consists of a set of operations and a combination of control structures that specify how the operations are to be arranged and executed. Similar to algorithms in the traditional sense, some human computation algorithms are more efficient than others. For example, if our computational problem is to map a set of images to tags, an efficient algorithm would make use of machine intelligence (e.g., active learning [284]) to select only images that the computer vision algorithm does not already know how to classify. Such an algorithm can greatly reduce the costs of the computation, both in terms of time and monetary payment to human workers. Some research questions relevant to the "what" aspect of human computation include the following.

- What tasks can be performed adequately by machines, therefore eliminating the need for human involvement? Can we leverage the complementary abilities of both humans and machines [149] to make computation more accurate and efficient?

- How do we decompose complex tasks into manageable units of computation and order them in such a way to handle the idiosyncrasy of human workers?

- How do we aggregate noisy and complex outputs from multiple human computers in the absence of ground truth?

The "Who" Aspect

Knowing what operations need to be performed, the next question is to whom each operation should be assigned. While for some tasks, aggregating the work of non-experts suffices, other tasks are knowledge intensive and require special expertise. For example, a doctor who is asked to verify that the fact "Obacillus Bordetella Pertussis is a bacterium" is likely a better (and faster) judge than someone without any medical training. Some research questions relevant to the "who" aspect of human computation include:

- What are some effective algorithms and interfaces (e.g., search or visualization) for routing tasks?

- How do we model the expertise of workers, which may be changing over time?

- What are some optimal strategies for allocating tasks to workers, if their availability, expertise, interests, competence and intents are known versus unknown?

The "How" Aspect

Finally, the "how" aspect pertains to the question of design—how can the system motivate workers to participate and to carry out the computational tasks to their best abilities (i.e., truthfully, accurately and efficiently). Some research questions relevant to the "how" aspect of human computation include:

- How do we motivate people to have a long-term interaction with the system, by creating an environment that meets their particular needs (e.g., to be entertained, to have a sense of accomplishment or to belong to a community)?

- How do we design game mechanisms [336] that incentivize workers to tell the truth, i.e., generate accurate outputs?

- What are some new markets, organizational structures or interaction models for defining how workers relate to each other (as opposed to working completely independently)?

The book is roughly organized by these three aspects of human computation, with Chapters 2 and 3 addressing the "what" aspect, Chapter 4 addressing the "who" aspect, and Chapters 5 and 6 addressing the "how" aspect.

1.3   TACKLING AI PROBLEMS: FROM VISION TO BIOLOGY

There are a variety of scientific and technological problems that are now benefiting from research in human computation.

1.3.1   A MATTER OF PERCEPTION

How do we recognize the objects that we see, or the instruments that we hear? What visual properties of images are beautiful to the human eye? What mood does music evoke? While the question of exactly how humans perceive is an important scientific one, our ability to build machines that can *mimic* human perception, i.e., understand visual scenes, identify objects and classify music, etc., can help address one of the most pressing technological problems today—the need to organize the billions of multimedia objects (images, music and videos) on the Web so that users can find the pins in the haystack with little effort.

Unfortunately, perception is a difficult problem for machines. For example, even with state of the art computer vision algorithms, machines are far from reaching human-level accuracy in object recognition. In the Large Scale Visual Recognition Challenge (LSVRC 2010) [155], a benchmarking competition for classifying 200,000 photos into 1,000 categories, the winning algorithm achieved 71.8% accuracy. Many computer vision algorithms are supervised learning methods, that are trained on a large number of examples (i.e., images) and their associated labels (i.e., tags). In LSRVRC, for example, a training set containing 1.2 million images was provided. Each of these images are processed by multiple workers [88] from the crowdsourcing platform Mechanical Turk, who receive small monetary payment for verifying the objects contained in each image. Another example of a project that involves human annotation of images is "The Last Tomb of Genghis Khan" project [316], sponsored by National Geographic, where volunteers help tag massive number of satellite images with the goal of resolving the age-old mystery about the location of the tomb of this ancient Mongolian emperor. As of 2010, the project has received over one millions tags, showing that the power of volunteerism cannot be under-estimated. In Chapter 6, we will introduce a different kind of human computation system, called *Games With A Purpose* (GWAP), that is capable of obtaining image labels from people for *free*, by engaging them in an activity (e.g., playing a fun game) that is intrinsically enjoyable to do. The most well-known example is the ESP Game [334] which has collected millions of image labels and was later adopted by Google (as the Google Image Labeler [130]) to help improve image search.

Audio perception is another difficult AI problem. Training a speech recognition system, for example, requires a large amount of audio transcription (i.e., speech-to-text mappings) data that has been traditionally costly to obtain [218]. There is also an exponentially growing amount of music data (e.g., music and sound clips) available over the Web; without efficient ways of automatically indexing this data, finding music on the Web can become a "pin in the haystack" ordeal. To address this problem, new machine learning methods [32, 143, 324] have been developed to automatically describe music and sound. There are many methods for obtaining labeled data for training these audio classifiers [323], from mining text (e.g., music reviews) from the web, paying volunteers, to building human computation games [24, 190, 215, 325]. TagATune, for example, is a human computation game that has collected over a million annotations from players, resulting in one of the largest labeled music datasets [212] publicly available to MIR researchers. Finally, the *evaluation* of algorithms in Music Information Retrieval (MIR) also requires substantial human effort. MIREX [230] is an annual benchmarking competition for evaluating and comparing MIR algorithms, e.g., for classifying audio, measuring music similarity, detecting onsets and keys, and retrieving music via a variety of modalities. In some cases, human judges were needed to evaluate the output of the algorithms. For example, to evaluate the algorithms submitted for the "music similarity and retrieval" track, typically around 40–50 researchers are required to each invest 3–4 hours of their time to help evaluate the competing algorithms. For evaluating learning algorithms, human computation games have also been shown [192] to be a viable alternative for collecting human judgments quickly and cheaply.

1.3.2   THE LANGUAGE BARRIER

A great deal of human knowledge, ideologies, beliefs and preferences are captured in written languages. As such, the building of machines to automatically parse and understand natural languages has been a central endeavor for researchers in Artificial Intelligence (AI), Information Retrieval (IR) and Natural Language Processing (NLP). Understanding languages can help us retrieve information on the Web relevant to free-form search queries, translate one language to another, extract facts, answer questions automatically and make accurate political, economic and cultural predictions about our world.

Machine learning algorithms have been applied to many natural language processing problems, including text classification, part of speech tagging, parsing, named entity recognition, word sense disambiguation, etc., each requiring a large number of labeled examples (e.g., documents and their classes, words in the context and their senses) as training data. For information retrieval, large scale benchmarking evaluations such as TREC (Text REtrieval Conference) [317] use datasets that were painstakingly hand labeled by experts. Statistical machine translation needs, as training data, large corpora of "parallel text" that map sentences from the source language to the target language. Such corpora are difficult to find, and the few that exist and are widely used, e.g., the European Parliament Proceedings Parallel Corpus (Europarl), are domain limited. For rarer languages, there is even less chance of finding adequate amount of training data for training machine translation algorithms.

Mechanical Turk now provides a relatively cheap medium for obtaining label data for NLP research [299]. However, with the use of this new platform come additional challenges. Annotation of text involves varying amount of effort and cost due to the inherent difficulty of the task or different capabilities of annotators [285], and both of these hidden variables can be difficult to estimate. To minimize cost, there has been several work on *active annotation*, the selection of annotation tasks to distribute to multiple workers [12] whose costs that are varied and unknown ahead of time. The idea of actively selecting computational tasks (Chapter 2) and assigning them to the *right* individuals (Chapter 4) to minimize cost and maximize quality is a central problem in human computation.

1.3.3   INTUITION INTO COMPUTATIONALLY INTENSIVE PROBLEMS

Despite the processing power of today's computers, many problems require computation that is of orders of magnitude beyond what computers can handle. Games (e.g., Chess [195] and Go [42]) and puzzles (e.g., Sudoku [209]) are prime example of problems that are computationally intensive for computers, but quite well mastered by humans. Likewise, there exists many NP-complete problems [123], such as the Traveling Salesman Problem [132, 140, 211], Packing [73] and FPGA [312], that cannot be solved with reasonable (i.e., polynomial) computational cost. Arguably, part of the reason is that we do not yet understand the human intuition that goes into solving these problems, and as a result, many algorithms resort to a "brute force" approach that exhaustively searches the space of solutions – a strategy which is feasible for small search spaces, but prohibitive when the problem instance is large. These difficult problems have been a central interest to AI research.

There has been several attempts to use human computation to provide a window into new solutions to these difficult problems. One such problem is the graph coloring problem, where the goal is to specify a color for each vertex of a network such that each node has a different color from its neighbors. The problem is notoriously difficult, and intractable even under weak assumptions [172]. A study was undertaken by [172] to learn how human participants can solve the graph coloring problem in a distributed way. Likewise, Corney et al. [73] studied how humans solve the packing problem, which is known to be NP-hard.

Other computationally intensive problems are found in the domain of biology, such as protein folding and genome sequencing. A protein is made of a chain of amino acids that folds into a particular shape (typically the lowest energy conformation), which determines the protein's function. Because of the large degree of freedom with which the chain of amino acids can move, the search space for the lowest energy conformation is astronomical, making the prediction of protein structure and function an extremely computationally intensive problem. To this end, a human computation game called FoldIt [69] was created, where human players are asked to re-fold proteins that have been inappropriately folded. It was found that the majority of players (even those with little scientific background) were capable of improving the protein folding solutions substantially. Moreover, results confirm that human computation can help advance AI by capturing what makes humans good problem solvers. For example, it was found that players were able to persist in backtracking, i.e., taking steps that seemed counter-productive but which eventually led to better solutions, whereas the computer program favored smaller adjustments which ended up generating more conservative solutions [69]. Following the success of FoldIt, another game was launched by McGill University to tackle the genetic sequencing [242] problem. The game Phylo [255], released in 2010, asks players to align colored blocks (representing the genetic code A, T, C, G) by moving them horizontally and by inserting gaps. The idea is to learn the strategies that people use in aligning sequences to help provide new insight into the multiple sequence alignment problem.

The key to solving computationally intensive search problems is a set of inventive heuristics for optimization [180, 181, 249]. This suggests a hybrid approach, with humans providing heuristics and high-level guidance and computers performing brute force computation in the much reduced search space. For example, Klaus et al. [180, 181] asked humans to specify the constraints (e.g., whether an item can be moved or not) in a sorting problem, thereby using humans to reduce the search space for machines. It was found that search time can be dramatically decreased with human in the loop.

1.4 OVERVIEW

This book attempts to define human computation as a research area, provide a comprehensive review of existing work, draw connections to a wide variety of disciplines and capturing their unique perspectives on the core research questions, and suggesting promising research directions in the field. To arrive at a set of fundamental, core research questions for human computation, we start with the following condensed definition of human computation:

> "Given a **computational problem** from a **requester**, design a **solution** using both **automated computers** and **human computers**."

Following this definition, the five most fundamental questions in human computation are the following:

1. What computational problems can or should be solved using human computers?

2. What are different types of solutions to a computational problem? For more explicit solutions, what are some programming paradigms for designing algorithms with humans in the loop?

3. For a requester, how do we guarantee that the solution is accurate, efficient and economical?

4. For a worker, how do we design an environment that motivates their participation and leverage their unique expertise and interests?

5. How do we leverage the joint efforts of both automated and human computers as workers, trading off each of their particular strengths and weaknesses? What other roles does machine intelligence play in human computation?

These questions are inherently inter-disciplinary, and for many of these questions, there already exists an extensive body of work in other research areas that partially address them. For example, the question of how to motivate workers to perform computation can be answered by studying how communities form and grow, how to design efficient markets so that tasks are priced appropriately, or how to infer workers' interests and expertise using machine learning methods. The desire to capture these rich connections is partly what motivates the writing of this book.

This book is divided into two parts. Part I focuses on the algorithmic side of human computation systems, i.e., how to design explicit procedures for finding solutions to a given computational problem. In Chapter 2, we will discuss human computation algorithms and how they are analogous to and diverge from algorithms in the traditional sense (i.e., without humans in the loop). In Chapter

3, we will discuss techniques for aggregating the noisy and complex outputs from multiple workers in the absence of ground truth. In Chapter 4, we will explain *task routing* – a component of a human computation system that chooses to *whom* to assign each computation task. Part II focuses on the design aspect of human computation systems. In Chapter 5, we will discuss different types of human computation markets and the characteristics of the workers within them, as well as ways to retain workers and motivate their continued participation. In addition, we will also discuss the needs of the requesters, and technology that can help address those needs. In Chapter 6, we will outline the important considerations when designing tasks, and review existing game mechanisms for eliciting truthful outputs from workers. Finally, in Chapter 7, we will provide a brief recap and suggest some future research directions for human computation.

PART I

Solving Computational Problems

CHAPTER 2

Human Computation Algorithms

Computation is the process of mapping some input representation to some output representation. By this definition, many everyday tasks can be called computational problems (e.g., consider Table 2.1 for some examples). Typically, in computer science, we study how to design *algorithms* to automatically solve these computational problems. For example, with the invention of automated computers, multiplications can now be trivially solved by computers. Likewise, sophisticated algorithms have been designed for automatically determining the condition of a patient, recognizing objects, translating text from one language to another, sorting quantifiable objects, and removing spelling and grammar mistakes from documents. Despite their availability, many of these automated algorithms are still far from achieving human-level performance. Other algorithms, such as sorting, cease to work well when the input objects are not numbers, but entities that require perceptual judgments and comparisons (e.g., sorting webpages by trustworthiness). In these cases, it is beneficial to involve humans in the loop to facilitate parts of the computation that are still too difficult for machines.

Table 2.1: Computational Problems		
Problem	**Input**	**Output**
multiplication	set of numbers	a number representing the product
sorting	set of objects	the set of object re-arranged in a particular order
medical diagnosis	x-ray and test results	cancer or no cancer
object recognition	image	a set of tags describing objects in the image
translation	sentence in language x	sentence in language y
editing	text	edited text
planning	goals and constraints	sequence of actions

As we move away from the traditional computational framework where all processes are automated, it is tempting to think of human computation as being very different. In contrast, we advocate the benefit of grounding the idea of *human computation algorithms* in the same language used in the study of algorithms in computer science, in order to borrow the well-established concepts (e.g., operations and control structures), programming paradigms (e.g., divide and conquer) and properties (e.g., correctness and efficiency) associated with traditional algorithms. In this chapter, we seek to explain human computation algorithms, in terms of how they are analogous to and diverge from traditional automated algorithms and review the existing tools for programming human computation algorithms.

2.1   A DEFINITION OF ALGORITHMS

Like the word "computation," the exact definition of the word "algorithm" has been under constant debate. Here, we adopt a particular definition of algorithms; this definition is not meant to be final, but rather it serves as a starting point for understanding the similarities and differences between algorithms with humans in the loop and without. It is expected that this definition will be modified and refined, as the field matures.

According to Knuth [182], an algorithm is "a finite set of rules which gives a sequence of operations for solving a specific type of problem," with five important properties.

- **Input**. The algorithm has one or more inputs.

- **Output**. The algorithm has one or more outputs, which have a specified relation to the input(s).

- **Finiteness**. The algorithm must always terminate after a finite number of steps.

- **Effectiveness**. Each operation of the algorithm needs to be sufficiently basic that they can in principle be done exactly and in a finite length of time by a man using paper and pencil.

- **Definiteness**. Each step of the algorithm must be precisely defined and unambiguous.

The first two properties are unquestionably part of human computation algorithms; in order for a process to be called an algorithm, there must be one or more *inputs* to the algorithm and one or more *outputs* that bear a specified relation to the inputs. Likewise, the properties of *finiteness* is particularly useful in distinguishing human computation from other social computing paradigms. In particular, in a human computation system, there should exist some processes (controlled and executed entirely or partially by humans) that are responsible for determining the answer after a finite (and reasonably small) number of steps; whereas in open crowdsourcing platforms such as Wikipedia, such explicit control processes do not typically exist.

The property *effectiveness* specifies that each operation must be sufficiently "basic" so that it can be completed within a finite length of time. A more concise definition is given by Stone [303], who defines an *effective* operation as a set of rules that a person can follow in a *robot-like manner*, without a need for thoughts. In designing an effective operation (e.g., to solve the quadratic equation), one must account for the fact that a person might not be equipped with the knowledge to solve the particular problem (e.g., they do not know how to extract a square root), in which case a set of rules must be included in the operation for the person to follow. As another example, suppose a human computer is asked whether the fact "Mario Lemeux is a hockey player" is true or not, they can be given some instructions such as "search for Mario Lemeux on the Web and find sentences that verify or disprove the fact." Instructions that accompany each operation, therefore, play a huge role in determining whether the operation is in fact effective and consequently whether the process is or is not an algorithm.

Definiteness is a desirable properties for human computation algorithms: it ensures that the output of the algorithm is repeatable. In practice, however, this property is hard to guarantee in a human computation algorithm. One way is reduce ambiguity to decompose complex operations into a set of simpler, more well-defined operations. For example, the "edit" operation on a sentence can be replaced by a combination of three operations—"find" (which finds a mistake), "fix" (which fixes the mistake) and "verify" (which verifies the fixes) [31]. The extent to which an operation should be decomposed, however, is more an art than a science. In other cases, it may be simply undesirable or impossible to specify precisely how an operation is to be performed. For example, it is easy to ask a human computer to find suspicious activities in images captured by the security video camera, but it would be difficult to enumerate a list of precise steps to detect all suspicious activities. Even if the list of precise steps are known, it is sometimes difficult to articulate them to humans in a such a way that nothing is left to interpretation.

2.2   BUILDING BLOCKS OF ALGORITHMS

2.2.1   OPERATIONS, CONTROLS AND PROGRAM SYNTHESIS

There are two main building blocks to an algorithm—**operations** and **controls** [183]. Operations are basic computational tasks with inputs and outputs. Controls (shown in Figure 2.1) are instructions for specifying how these basic operations (e.g., in what order, how many times, etc.) are to be executed. According to Boehm and Jocipini [37], there are three main types of controls: *sequence (or iteration)*, which specifies the order in which operations are executed, *selection (or choice)*, which specifies a set of conditions and the specific operation to be executed when each condition is met, and *repetition (or looping)*, which specifies what operations to execute repeatedly and the condition for terminating the loop. In parallel or distributed computing [18, 297], another important control structure is *parallel*, which specifies a set of operations to be executed simultaneously, after which their individual outputs are aggregated. Any algorithms can be considered as an arbitrary combination of these control structures, which together outline the exact steps needed to solve the given computational problem.

Many existing human computation systems can be described as a combination of these building blocks. For example, the ESP Game is essentially a parallel algorithm that simultaneously distributes multiple labeling tasks for each image until it is considered successfully labeled (e.g., when enough high confidence tags are collected). A more complicated example is QuickSort [71], which involves a mixture of sequence, parallel, selection and repetition controls. Given an array, QuickSort executes two operations in sequence—pivot (which involves picking an item in a list) and sort (which involves dividing the array into two smaller arrays, each containing items with lesser and greater values than the pivot respectively). The algorithm then repeats the process until the array cannot be decomposed further. By involving humans in the loop to perform the pivot and sort operations, QuickSort can be adapted to perform subjective sorting [204], i.e., ranking objects by some attribute that is difficult to quantify by a machine (e.g., attractiveness) and for which well-calibrated rating

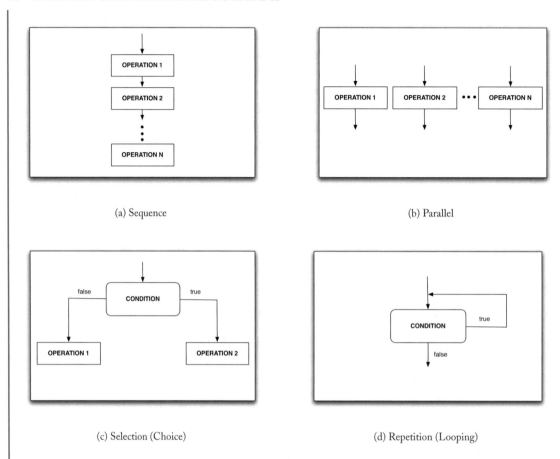

(a) Sequence   (b) Parallel

(c) Selection (Choice)   (d) Repetition (Looping)

Figure 2.1: Control structures.

scales do not exist [55, 133]. As another example, consider the human version of a distributed algorithm for solving the graph coloring problem [172, 216]—assigning colors to the vertices of a graph such that each vertex is colored differently from its neighbors. Kearns et al. proposes a human computation algorithm [172], where each worker is given a local view of the graph (e.g., the color of a particular node and of its neighboring nodes), and asked to update the color of his node to resolve any conflicts. It was found that across a variety of graph structures, people were able to come up with the optimal solution within the allotted time period. In each of these examples, the control structure is pre-defined and executed by a machine, with the basic operations carried out by humans.

Instead of just performing the operations, humans can also handle the control flow of the algorithm, by making decisions that are difficult for machines to make. For example, human workers can decide whether the program should halt by judging whether the current solution is "good enough." They can make conditional decisions throughout the execution of the program, e.g., decide whether a complex operation should be decomposed into simpler ones or not [185], whether to backtrack (e.g., if they detect a problem in the current solution), or which operation to execute next. The design of the control flow—deciding how many times to iterate on a task, how many votes it takes to determine the consensus opinion—can also be determined automatically. For example, TurKontrol [77, 78] learns a Partially Observable Markov Decision Process (POMDP) model that automatically determines which operation to execute next based on the probability that an average worker can improve upon the current solution.

Beyond performing operations and controls, human workers can put together an entire algorithm from scratch by creating new operations and controls on the fly, a process called *program synthesis* [362]. For example, given a computational problem (e.g., write an essay about topic *X*), Turkomatic [185] coordinates workers to come up with an algorithm to solve the problem; it does so by running a meta-algorithm that consists of three operations—*solve* (which asks the worker to perform a task), *decompose* (which asks the worker to break down a task into two or more subtasks) and *combine* (which asks the worker to aggregate the solutions from a set of subtasks). The controls of this meta-algorithm—e.g., the decision of whether to solve or decompose—is also determined by the workers. In other words, the process of program synthesis is itself governed by an algorithm, i.e., there are explicit "solve," "decompose" and "combine" operations that workers have to follow to create the algorithm for solving the computational problem at hand. Having workers perform program synthesis essentially eliminates the need to design algorithms for every new problems that come along, and also leaves the burden of design decisions, such as what operations to use, how granular each operation is, or how many workers to assign to each operation, etc., entirely up to the human computers. A similar paradigm is proposed by CrowdForge [178].

Typically, each operation, control or synthesis step is mapped to a task, which can take a variety of formats, such as a round of a human computation game or a HIT (Human Intelligence Task) on Mechanical Turk. The particular design of the task can directly affect the quality of the final output. Design considerations include how instructions are presented, whether to include sample or partial solutions, how much to pay workers so that the reward is appropriate for the level of effort, and how to prevent or minimize cheating or spam, etc. In Part II of this book, we will review different strategies for designing tasks for human computers that help to ensure that the algorithm, upon execution, can actually produce high quality outputs.

2.3   PROGRAMMING FRAMEWORKS

There are a few toolkits that provide programmatic access to human workers on Amazon Mechanical Turk (called *Turkers*), including TurKit [204, 322] and Boto [41]. These toolkits offer wrapper methods in different programming languages (e.g., Java, Python) for calling Amazon's MTurk REST API. Using these toolkits, requesters can write computer program as they normally would, but in addition, they are given the ability to post operations to be carried out by Turkers and retrieve the results.

```
/* this function creates a recipe (a list of steps) that include the given
list of ingredients and a recipe name */

function create_recipe(ingredients, recipe_name)
{
    var steps = array();

    // gather a list of steps for making the recipe
    while (!is_done(recipe_name, ingredients, steps)) <-- human-driven condition
    {
        new_step = get_next_step(recipe_name, ingredients, steps); <-- human-driven operation
        steps.add(new_step);
    }

    return steps;
}

function is_done(recipe_name, ingredients, steps)
{
    var hitID = once createHIT( ... ''is_done.html'' ... );
    var answer = once mturk.waitForHIT(hitID)
    return h.answer[0];
}

function get_next_step(recipe_name, ingredients, steps)
{
    var hitID = once createHIT( ... ''get_next_step.html'' ... );
    var answer = once mturk.waitForHIT(hitID)
    return h.answer[0];
}
```

Figure 2.2: A human computation algorithm for creating recipes.

Getting humans to perform tasks is expensive—it costs money and takes time, especially factoring in the time it takes for Turkers to *discover* the tasks. Therefore, any human computation programming framework must provide ways to manage cost and recover from execution error [204]. For example, TurKit [204] provides functionality to preserve the state of the program, by storing the results of previously executed HITs in a database. If a program crashes, then the algorithm can be re-run without losing information and having to re-post any previously executed tasks.

Consider the pseudocode in Figure 2.2 showing a simple human computation algorithm for creating a recipe (a list of steps) for cooking a particular dish, given as input a list of ingredients and the name of the recipe. This example shows that humans can be involved in performing basic operations (e.g., creating the next step in the recipe) as well as condition checking (e.g., ensuring that all the ingredients are cooked). The built in functions *createHIT* and *waitForHIT* are used to post a task onto Mechanical Turk and retrieve the results. The primitive **once** ensures that the result of the HIT is stored, eliminating the need to re-execute the same HIT if the program crashes. Finally, parallelization of operations can be achieved through the **join/fork** primitive. More details about the Turkit can be found in [87, 322].

As it was pointed out in [139], future programming environments should provide access to a variety of human computation markets (e.g., Mechanical Turk, games with a purpose, oDesk, etc.), to post tasks, retrieve results, approve and reject completed work; more sophisticated environments may even provide automated service for recruiting workers from any marketplaces, eliminating the needs for developers to worry about anything else except about the program logic of their algorithms.

2.4   EVALUATING HUMAN COMPUTATION ALGORITHMS

Until now, there has been little research on formally characterizing the *quality* of a human computation algorithm. For example, given two different human computation algorithms for sorting, how do we know which is better? In the traditional study of algorithms, there are two main criteria for judging the merits of algorithms—correctness and efficiency.

- **Correctness**. Does the algorithm lead to a solution of the problem in a finite number of steps?

- **Efficiency**. How much resources (in terms of memory and time) does it take to execute the algorithm (in the size of the input space)? Is the algorithm cost-effective? (i.e., how much monetary cost will be incurred)?

In this section, we will discuss whether human computation algorithms can be evaluated using the same criteria as traditional algorithms. With the exception of Shahaf et al. [287], there has been very little work on formalizing the notion of complexity in human computation.

2.4.1   CORRECTNESS

To check for correctness of automated algorithms, specifications are commonly used—one defines the pre-conditions (the state of the variables before the execution of the program) and the expected post-conditions (the state of the variables after the execution of the program), and compares the two in order to establish correctness. When computation involves humans in the loop, however, the outputs of any of the operations can be noisy, either due to mistakes or variability in judgments and opinions. An alternative way to compare human computation algorithms, therefore, may involve evaluating the *robustness* of the algorithms in producing the correct answer in the presence of noise [103, 257, 340]. It remains an important open question whether it is possible to characterize the correctness of human computation algorithms formally.

2.4.2   EFFICIENCY

Efficiency has a much more straightforward and analogous interpretation. It is, in fact, one of the most important concerns in human computation. Typically, a given computational problem needs to be solved within a certain time limit [226, 227] (e.g., the translation of a news article within one week). In some applications, there is even a need for real-time computation. For example, Viswiz [35] is a human computation system that allows blind individuals to upload images from a mobile phone, and receive a description of the image in (near) real time. In this case, the system must ensure that the answer comes back quickly, in spite of the fact that the availability and efficiency of workers may be unknown. There are many ways to characterize the efficiency of an algorithm, including: (i) **time complexity**, which can be measured in terms of run-time complexity (i.e., the number of operations) and clock time (i.e., the actual amount of time it takes for all the operations to execute); (ii) **query complexity**, which is the total number of queries to human oracles; and (iii) **cost-effectiveness**, which is total amount of monetary costs incurred. Out of these three properties, cost-effectiveness is an important one that is not typically used to characterize automated algorithms.

2.4.2.1 Time Complexity

There are two notions of time complexity. Run-time complexity concerns how the number of operations scales as the input size of the algorithm increases. This is particularly important for cost management—an algorithm that does not scale (e.g., the number of operations increases exponentially in terms of the input size) can incur astronomical monetary costs when the operations are carried out by paid workers.

Clock time is the actual amount of time spent on the computation, which is tied to how efficiently individual human computers can perform their given tasks. If a task is too ambiguous, complex, de-motivating or cognitively overwhelming, then the total amount of time it takes to run the human computation algorithm can be affected. In other words, the design of a task—from granularity to the clarity of the provided information—can directly affect the efficiency of the algorithm. An additional factor to take into account is the *discovery time*—the time it takes for workers to find the tasks. Discovery time is affected by the effectiveness of the particular search interfaces [64], recruitment [30] and pricing strategies [108, 295].

2.4.2.2 Query Complexity

Query complexity measures the number of queries to human oracles it takes to process all the input objects. In human computation, query complexity is affected by two decisions that an algorithm can make—for a given input object, (1) whether it should be processed by a human worker or by a machine, and (2) whether it should be repeatedly processed by multiple workers.

The first decision is a typical one made by active learning algorithms. **Active learning** [245, 284] (or active annotation [20, 331]) is a machine learning model in which the learner is given the opportunity to choose the data from which it learns, and the goal is to learn the function of interest by asking as few questions (i.e., to a fictitious *oracle*) as possible. This is in part motivated by the fact that for many computational problems, a large amount of labeled data is needed to train automated algorithms, and obtaining such labeled data is costly. Consider the problem of image classification. The naïve human computation algorithm would distribute every image to multiple workers. A more efficient algorithm would select images to query whose labels provide non-redundant information for improving the current model, then use the model to label other images automatically. There are also several variations of active learning algorithms [284]—in membership query synthesis, the learner can make a query about a synthesized example; in stream-based selective sampling, unlabeled examples is sampled one at a time from a distribution, and the learner decides whether to label or discard each example; finally, in pool-based selective sampling, unlabeled examples are gathered all at once, and the learner can evaluate all the examples in this pool and greedily select the ones that are the most informative.

Much of the research in active learning is centered around the design of *query strategies*, i.e., how to select examples to query in order to minimize query complexity. Some common query strategies include uncertainty sampling [196], which samples examples whose labels are the most uncertain, query-by-committee [286], which samples examples whose labels are disagreed upon by the most experts, or expected error reduction [276], which samples examples whose labels are expected to have the most effects on improving the generalization power of the classifier. Finally, while there has been some work on using humans as oracles [263], active learning typically assumes a hypothetical existence of a single, perfect, omniscient oracle. This assumption breaks down as we move towards a framework with human computers as oracles. There are now new active learning frameworks [93, 341] that bridge the reality by assuming that there exist multiple oracles that can be imperfect, unreliable, reluctant to answer queries, and distinct in their domains and levels of expertise.

The second decision addresses the question of **repeated labeling** [290]—since there are multiple imperfect oracles, the algorithm should sometimes make multiple queries for the same input object, in order to be confident about the obtained label. Repeated labeling can lead to higher quality solutions [290] but also incurs additional costs. Human computation algorithms that intelligently decide when and how much to repeatedly label are potentially more efficient.

2.4.2.3 Cost-Effectiveness

This is related to efficiency, but involves not only choosing which input objects to process but also the appropriate strategies for pricing tasks. In many situations, requesters are limited by budget constraints [205, 344] (e.g., they may only be willing to pay $100), in which case the system might need to choose tasks that are the most *cost-effective* [330]. In other words, algorithms can be compared in terms of how well they can simultaneously satisfy a requester's cost and quality requirements. Machine intelligence can be of help here. For example, there has been some work on *automated task design* that determines parameters of a task, e.g., its granularity and pricing, automatically [150].

2.5 SUMMARY

This chapter addresses the first fundamental problem—how to design an algorithm for solving a computational problem using human computers. The key points of this chapter includes the following:

- There are ways in which human computation algorithms are analogous to or diverge from automated algorithms.

- Correctness and efficiency are two useful criteria for evaluating and comparing human computation algorithms.

- Machine intelligence can be used to automate and improve the design of human computation algorithms.

CHAPTER 3

Aggregating Outputs

A common practice in human computation is to have multiple human computers perform the same operations. For example, in human computation games, multiple pairs of players are asked to generate tags for any given object (e.g., image [334], music clip [190] and named entities [338]). There are many reasons why redundancy is valuable. First, the data obtained from human workers can be inaccurate: some human workers may give the wrong answers due to the lack of expertise, or suboptimal physical and psychological condition at the time of computation. Second, truth-elicitation mechanisms (see Chapter 6) are rarely, if at all, used in human computation systems other than games, because they require the synchronization of multiple workers, a functionality that is not available in most crowdsourcing marketplaces (e.g., Mechanical Turk). Third, even if these game mechanisms are in place, there still exists the possibility of malicious workers finding a way to cheat or spam the system. Finally, even if all human workers are truthful and accurate, substantial variation in their answers can still arise due to individual differences in perception or interpretation of the question. A piece of music that has been labeled by some as "angry" may be perceived as "energetic" by others; the same sentence in one language can be translated into many semantically equivalent sentences in another language. Since the ground truth is unknown, the challenge then becomes knowing which outputs to trust, which to discard, and how to take two outputs and automatically merge them together to form a single output.

There exist both human-driven and automated techniques for aggregating the outputs of multiple human computers. In this chapter, we review automated techniques for aggregating outputs, focusing on two different classes of problems: (1) aggregation of simple outputs, e.g., multiple class labels for the same input object; and (2) aggregation of complex or structured outputs, e.g., rankings, clusterings, sentences and beliefs. Human-driven techniques (e.g., manual verification, voting, filtering, and merging) are left for discussion in Chapter 6. Automated techniques have the advantage that they are cost-effective and scalable, and they vary in their assumptions about the process through which output data were generated by human computers, e.g., whether the input objects vary in difficulty, or whether human computers have different competence and expertise that affects their abilities to produce the correct output. They also share a common assumption: what we are trying to compute has a true answer, and that answer can be approximated using the redundant responses of multiple human computers.

3.1   OBJECTIVE VERSUS CULTURAL TRUTH

Before describing any techniques, it is worth distinguishing the two types of functions that can be computed through human computation. The first type of functions have outputs that are *objectively* true, i.e., there exists a definitive answer that is external to human judgment. An example of such functions is the mapping of patients to diagnoses—the question of "whether a patient has cancer or not" can be determined definitively by performing a biopsy. Where human computation becomes useful is when the true output cannot be determined definitively, due to factors such as the lack of resources or imprecision in the measuring instruments. For example, biopsies are expensive or simply impossible to perform; in this case, the answer is often determined by aggregating the opinions of several medical experts who review the patient's case. Other examples include having a set of border control staff determine from an X-ray image whether a suitcase contains dangerous items or not, or having a group of geologists identify volcanos by inspecting images of Venus [298]. In both cases, the truth is external to and unaltered by human judgments.

The second type of functions have outputs that are considered "culturally" true, i.e., the true answer refers to the shared beliefs amongst the set of people that we sample, and determining this answer usually involves some sort of perceptual judgment. Examples include mapping a piece of music to the emotions (e.g., "serenity") that it evokes, determining whether a particular website contains pornographic content or not, and rating the attractiveness of a celebrity. In these cases, it is difficult to derive an objective, true answer. This is because "serenity," "pornography" and "attractiveness" are perceptual concepts that are difficult to define precisely, and their interpretation can vary greatly from culture to culture, or even from individuals to individuals. For example, mothers may be more conservative than college students in judging whether a website is pornographic or not [158], and what facial features are considered beautiful can vary depend on the cultural background of the individual annotators.

In the absence of ground truth, we rely on the "wisdom of the crowd" to establish what is truth. In the former case, we assume that there is a hidden objective truth that can be approximated by aggregating the outputs of many workers, given that at least some workers are accurate. In the latter case, we assume that there is a cultural consensus amongst the sample of workers that can be identified even if some workers may deviate from the norm. That said, if every worker lies, then there will be little hope for discovering the truth by aggregating redundant outputs.

3.2   CLASSIFICATION

In this section, we focus on the problem of *classification*, where workers are asked to put a set of objects (e.g., images, music, patient records) into a fixed number of categories.

3.2.1   LATENT CLASS MODELS

For classification, most automated output aggregation methods, even simple techniques such as majority votes, are based on what is called *latent class models*. These generative models typically assume that what is *observed* are the outputs that human computers generated for a given operation, and what is *latent* (i.e., hidden) are the ground truth and other factors, such as worker's competence and the difficulty of the task, that influenced how the observed outputs were generated. More formally, suppose there are N computational tasks, where the *true* output Y_n for each task is unknown. Our goal is to estimate Y_n given an output matrix O below, containing the responses from M workers. In most situations, the output matrix will have missing values, since we may not be able to find M unique workers to perform all of the tasks.

$$
\begin{array}{c}
\text{computational task} \\
\begin{array}{cccc} 1 & 2 & \cdots & N \end{array} \\
\text{worker}\ \begin{array}{c} 1 \\ 2 \\ \vdots \\ M \end{array}
\left[\begin{array}{cccc}
O_{11} & O_{12} & \cdots & O_{1N} \\
O_{21} & O_{22} & \cdots & O_{2N} \\
\vdots & \vdots & \ddots & \vdots \\
O_{M1} & O_{M2} & \cdots & O_{MN}
\end{array} \right]
\end{array}
$$

This type of output matrices exist in many real world problems. In medicine, multiple doctors may give their expert opinions on whether their patients are diseased or not based on X-ray images [349]. Likewise, anthropologists often ask questions to probe the beliefs held by members of a culture [271], without knowing the true answers nor the competence of the individuals ahead of time. For example, it is known that members of a culture share beliefs about which diseases are contagious versus not [270, 347], and that some individuals (e.g., those with many children) are better at giving the culturally correct answer. In these scenarios, the ground truth is unknown and we must infer the true answer from the noisy judgments of multiple people.

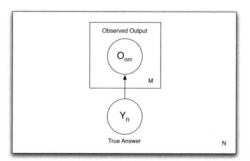

Figure 3.1: Majority vote.

The simplest aggregation method is *majority vote*, which can be represented as a simple probabilistic model (see Figure 3.1) that makes little assumptions about the annotation process, except that the output that each worker independently generates depends on the true answer, and there is no prior information about which categories are more or less likely to be the true classification, i.e., $P(Y_n = j) = 1/J$. More formally, given the output matrix O, our goal is to determine for each computation task n the category that is most likely the true answer, that is:

$$Y_n = \operatorname*{argmax}_{j} P(Y_n = j|O) \tag{3.1}$$

$$= \operatorname*{argmax}_{j} \frac{\prod_{m=1}^{M} P(O_{n,m} = o_{n,m}|Y_n = j)P(Y_n = j)}{P(O)} \tag{3.2}$$

$$\propto \operatorname*{argmax}_{j} \prod_{m=1}^{M} P(O_{n,m} = o_{n,m}|Y_n = j) \tag{3.3}$$

$$\propto \operatorname*{argmax}_{j} (1 - \epsilon)^{\sum_{m=1}^{M} \mathbf{1}(o_{n,m}=j)} \cdot \epsilon^{\sum_{m=1}^{M} \mathbf{1}(o_{n,m}\neq j)} \tag{3.4}$$

The problem of inferring the ground truth, therefore, is equivalent to the estimation of data likelihood $P(O_{n,m} = o_{n,m}|Y_n = j)$. Suppose that we assume that each worker outputs the correct answer with the same fixed probability $1 - \epsilon$, and that ϵ is strictly less than $1/2$, then the model essentially says that the more workers who *agree* that the true answer is j, the more likely the true answer is in fact j.

Using majority vote to identify the correct answer has a few shortcomings. It does not take into account the fact that workers can make random guesses or make mistakes and still agree by chance [13]. This is especially problematic if the majority of the workers are novices (who systematically make the same kinds of errors) or spammers (who generate answers at random). Additionally, many of the factors that influence the outcome of the computation are not captured by the simple model. First, each worker may have different *biases*. For example, when asked to rate webpages on a scale in terms of adult content, workers who are mothers may be much more conservative than college students [158]. Human computers may also have different *expertise* [52, 345]. Bird watchers who live by the coast may be much more capable at distinguishing between different seabird species than those who live far from the water. Finally, the quality of workers' outputs can be affected by the inherent *difficulty* of the classification task. For example, for an ambiguous image, regardless of their expertise, *all* workers may confuse its classification amongst different categories.

In other words, the output of workers depends not only on the true answer of the computational tasks, but also a set of hidden variables θ, such as the general competence of the workers. One of the earliest examples of these more complex output aggregation models was proposed by Dawid and Skene ([79]), for classifying the unknown true states of health of patients (e.g., whether they are in good enough condition to undergo a general anaesthetic) given the assessments (e.g., ratings) of multiple doctors. Instead of all clinicians having the same fixed error rate ϵ, it is assumed that each clinician makes different types and amount of errors. Specifically, the error rates of each clinician k can be captured by a confusion matrix π, where $\pi_{jl}^{(k)}$ specifies how likely the clinician will declare a patient to be in a state l when the true state of the patient is in fact $j, l = 1, \cdots, J$ and $j = 1, \cdots, J$.

The idea behind the model of Dawid and Skene is to infer the confusion matrix and the true state of the patients simultaneously using the Expectation-Maximization (EM) algorithm, which iteratively (i) estimates the true states of each patient by weighing the votes of the clinicians according to our current estimates of their competence (as given by the confusion matrix), and (ii) re-estimates the confusion matrices based on the current beliefs about the true states of each patient. There are many other models [54, 86, 158, 264, 265, 327, 346, 350] that take into account the differing competence and biases of individuals, including a set of statistical models developed by anthropologists called Cultural Consensus Theory [171, 271, 348, 354], in which the culturally correct answers and the cultural competence of individuals are the hidden variables.

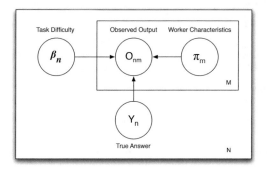

Figure 3.2: Latent class model.

More sophisticated latent class models can include other latent factors that affect the computational process, including biases, expertise, and task difficulty (Figure 3.2) [345, 357]. Welinder et al. [345], for example, incorporate into their bird image classification model factors such as: (a) each worker's competence which varies depending on the bird image; (b) each worker's biases, which is related to how they individually assign costs to different kinds of mistakes; and (c) the difficulty of the image itself (e.g., an image may have "nuisance" factors such as viewpoint differences, pose variations and occlusion, which make it difficult to classify). In adaptive testing, similar models are useful for simultaneously inferring the true answers to open-ended questions and the competence of the students [17, 252].

3.2.2   LEARNING FROM IMPERFECT DATA

There are two major assumptions made by latent class models—that the system can reliably distribute the same task to many *unique* workers, and that each worker would do enough tasks that his or her expertise/competence can be reliably estimated. Dekel et al. [83] introduces output aggregation techniques that will work when these assumptions do not hold. In the absence of repeated labeling, they propose to train a classifier on the set of unfiltered data collected from workers, then use the learned hypothesis as a proxy for ground truth and use it to prune away bad teachers whose labels do not agree with the learned hypothesis. As a result, low quality data can be removed even when some tasks are performed only once.

Even after aggregating and filtering outputs, there can still be a huge amount of noise in the data, which poses a challenge to applications (e.g., machine learning algorithms) that need to make use of it. There are two types of noise. The first type is *inaccurate labels* [44]. For this, there has been some work on learning methods that can withstand inaccurate labels with the assumption that these inaccuracies are due to multiple workers with varying levels of unreliability [82, 207, 264, 265, 268, 298].

CLASSICAL	GRUITAR	FEMALE	RENNAISSANCE	STOMP
GUITAR	PRIMAL	VOCAL	SWING	SKIPPY
PIANO	ACCUSTIC	QUIET	SCI-FI	FOREIGN
VIOLIN	ACTIVE	SITAR	HIPPIE	CHRISTMASSY
ROCK	MEOW	CLASSIC	LULLABY	CLAPPY
SLOW	OHOHOH	SOFT	ANGELIC	CLOUDY
STRINGS	GRADUAL	CELLO	DOWNBEAT	SEASIDE
TECHNO	CLIMATIC	WOMAN	RELAXATION	MAMBO
OPERA	PENSIVE	MALE	GLOOMY	MANDOLIN
DRUMS	HOUSY	SINGING	ROYAL	FOLK
SAME	INSTRUMENTAL	VOCALS	RYTHMIC	NO VIOLINS
FLUTE	CALMISH	SOLO	MUFFLED	MELODY
FAST	FEMALE OPERA	LOUD	RAGTIME	HARMONICA
DIFF	VARIED	CHOIR	TUDOR	ITALIAN
ELECTRONIC	HEALING	VIOLINS	FANTASY	DRAMATIC
AMBIENT	WAVEY	HARP	HISPANIC	BLUEGRASS
BEAT	DRIPPING	BEATS	BEATLES	GENTLE
HARPSICORD	HEBREW	NOT ROCK	SYNCOPATED	SPACESHIP DESCENDING
SYNTH	ANIMALS	WIERD	MID-TEMPO	COOKIE MONSTER VOCAL
INDIAN	REEDS	DANCE	RATTLE	VAMPIRES AT A DINNER PARTY

Figure 3.3: Tags collected by TagATune.

The second type is *redundant labels* due to the allowance of free-form, open vocabulary answers (see Figure 3.3 for an example). For example, in most human computation games (e.g., the ESP Game, TagATune), there are no restrictions on what tags players can enter. Consequently, the resulting datasets end up having a huge number of classes (i.e., many equivalent classes due to synonyms and spelling mistakes/variations) of imbalanced sizes (i.e., some having very few examples, others with lots of examples) [189]. Techniques for synonym resolution [26, 358] may help reduce this type of noise. In general, learning methods that are capable of learning directly from the *unprocessed* data collected by human computation systems is an interesting area for future research.

3.3   BEYOND CLASSIFICATION

There are many other computational problems, e.g., ranking and clustering, that have more complex output than simple labels. To involve humans in solving these computational problems, two questions need to be addressed. First, how should the problem be decomposed? For example, to rank a set of images, are the workers asked to vote for the best image amongst a subset, perform paired comparisons, or provide the complete ranking of all the images? Second, what are some techniques for aggregating the output of each operation and producing a consensus output? For example, how should we aggregate a set of partial rankings to generate a complete ranking that actually reflects the opinion of the crowd? To date, not many automated methods for aggregating complex outputs have been used in human computation algorithms, even though these techniques exist and have been applied in other settings. In this section, we will review the existing output aggregation techniques for ranking/voting, clustering, combining structured outputs (e.g., sentences) and beliefs.

3.3.1   RANKING AND VOTING

To rank is to order a set of objects by their values for a particular attribute. For some attributes, there is no objective scale and it is often much easier for people to express their judgments in terms of paired comparisons (e.g., $A > B$) than to articulate the actual attribute value for any given object. In other cases, the number of objects to rank can sometimes be huge, making it infeasible to ask each worker to provide a complete ranking of all the objects. Instead, the partial ranking for a subset is solicited. For example, the game Matchin [133] gives players a pair of images and ask them which one they prefer; these paired comparisons were then combined to form a single global ranking using three different aggregation methods: ELO (typically used to rank chess players) [101], TrueSkills [135] and a variant of SVD (Single Value Decomposition) [309]. Finally, even if the number of objects is manageable for a single worker to rank, one might want to aggregate the rankings provided by multiple individuals to form a *consensus ranking*, in order to eliminate errors and minimize personal biases. This problem of combining rankings is called the *rank aggregation* problem, and has been studied extensively [3, 106] for combining rankings that are generated by machines (e.g., search results rankings by multiple search engines [96]) or by humans (e.g., preference judgments obtained from clickthrough data [67, 164, 262]). Closely related to rank aggregation is the idea of vote aggregation in social choice theory [11, 65, 142] and rating aggregation [122, 125].

3.3.2   CLUSTERING

Consensus clustering [3, 127, 129, 304] (also known as cluster aggregation or cluster ensemble) is the process of producing a single clustering that minimizes the disagreement amongst multiple clusterings of the same objects. Among many advantages, consensus clustering can help improve robustness, by combining the clusters produced by multiple algorithms or human computers.

For a large set of objects, it is virtually impossible for a single worker to provide the complete clustering. Similar to the case of ranking, the clustering problem should be decomposed into simpler tasks that are easy for individual workers to do. For example, instead of specifying the clusters, workers can specify must-link (i.e., whether two objects should belong to the same cluster) and cannot-link (i.e., whether two objects should not belong to the same cluster) constraints [151] between pairs of objects in the set. Clustering algorithm can then use these constraints to recompute a clustering that satisfies as many constraints as possible.

3.3.3 STRUCTURED OUTPUTS

A common task on Mechanical Turk asks workers to generate descriptions of images. If multiple workers are asked to describe an image, then there needs to be some process in place for filtering out lesser descriptions [202] or merging multiple descriptions. In Natural Language Processing (NLP), there has been research on automatically combining similar text snippets from multiple sources in order to form a summary [25, 27]; these *information fusion* techniques can potentially be used to eliminate manual filtering and merging tasks, thereby lowering the cost of human computation substantially.

Machine translation is another example where multiple outputs (e.g., translated sentences) are typically elicited, and the human computation system must decide how to generate a single output translation, either by selecting the best sentence [8, 256, 360] or merging the results to derive a *consensus translation* [21, 116, 163, 222, 273].

3.3.4 BELIEFS

Prediction markets [61, 62, 254, 355], also known as "information markets" or "event futures," are a method for aggregating the beliefs of many individuals. The premise behind prediction markets is that knowledge is distributed, and by aggregating the information of many individuals, we can accurately forecast future events (e.g., election results or box office earnings) and estimate certain subjective (e.g., the quality of a product) or unmeasurable (e.g., whether the world will end) outcomes. In such markets, participants can buy and sell *contracts*, and the payoffs are determined by the outcome of the event. Participants are given the incentives to gather information and report their opinion truthfully, and the market aggregates the opinions and reveal some measure of the event outcome, e.g., in terms of the probability that an event will happen, the mean or median value of an event outcome, as well as the uncertainties about these expectations [355]. Substantial research goes into the design of prediction markets, including the mechanism of interaction between buyers and sellers, the specification of the contract, and whether real money is involved and supporting information is available [62, 355]. Given that many computational tasks of interests to human computation are inherently subjective, peer prediction markets [166, 229], which aggregate subjective assessments (e.g., movie ratings), may be a viable way to collect and aggregate beliefs-like outputs from human computers in next-generation human computation systems.

3.4 SUMMARY

In human computation, part of the solution to a computational problem involves aggregating the outputs of many human computers. In this chapter, we address the problem of aggregating noisy and complex outputs when the ground truth is unknown. The key points of this chapter include the following:

- To classify an object, a human computation algorithm must aggregate the labels produced by multiple individuals to produce a single class label. The most common method is by building a latent class model to model the way labels are generated by workers, taking into account factors such as task difficulty and worker competence.

- Although not as prevalent in current human computation systems, there exist many techniques for aggregating complex and structured outputs (e.g., rankings, clusterings, sentences and beliefs) that can be leveraged by human computation algorithms.

CHAPTER 4

Task Routing

One of the most powerful pieces of evidence that affirms the benefit of the crowd over individuals is the observation that the work of many non-experts can be aggregated to approximate the answer of an expert [8, 299]. However, as we have seen in the previous chapter, the competence and expertise of the workers *do* matter—outputs from different workers should be trusted to different extents depending on how reliable they are. This is especially important for tasks that are knowledge-intensive. For example, determining whether an X-ray image indicates cancer requires medical knowledge; determining whether a sound clip is from a Junco or a Wren is best answered by an avid birder than someone who has no knowledge of birds. Beyond being *domain-specific* (e.g., some people know more about sports while others know more about gardening), expertise can also depend on the characteristics of the tasks (e.g., some people are more detail oriented and prefer repetitive tasks, while others are more creative and prefer tasks with open-ended questions). Under the assumption that experts can handle knowledge-intensive tasks better, task routing is in fact another point of intervention for ensuring output quality *before* computation actually takes place—instead of assigning tasks to workers randomly and only learning what outputs to retain or discard *after the fact*, the system can be equipped with more intelligent ways of matching workers to tasks they can do best. We call this problem **task routing**.

Experts can have many advantages over non-experts. In the study of expertise [63], it has been shown that experts are better at generating better, faster and more accurate solutions, detecting features and deeper structures in problems, adding domain-specific and general constraints to problems, self monitoring and judging the difficulty of the task, choosing effective strategies, actively seeking information and resources to solve problems, and retrieving domain knowledge with little cognitive effort. On the other hand, expertise is not necessarily the only criterion for matching. For example, in order to promote *learning*, e.g., in Duolingo, the system should occasionally assign a translation task to novice workers, at the risk of receiving bad data, in order to build their skills at translation. In human computation games, the tasks chosen for players must be balanced for difficulty and fun, providing enough challenge yet not depriving players of their sense of accomplishment. The task routing problem takes many forms—it can mean finding an expert for a single task out of thousands of workers, or distributing a large number of tasks amongst a few workers. Consider the following scenarios as an indication of the diversity of task routing problems and solutions.

- "Paying Turkers to translate a large corpus of text from French to English sentence by sentence."
  The system (i.e., Mechanical Turk) can single out workers who are well versed in both languages by giving each worker a qualification test. However, the decision of which sentence to translate is left entirely up to each worker, who is given a simple search interface to find suitable tasks.

- "Hiring a small team of full-time workers to judge the relevance of webpages to thousands of queries."
  The system has the opportunity to learn about the preferences and abilities of workers over time, and push them the most suitable tasks. The challenge here is to automatically infer what the search queries mean and how they relate to each worker's profile.

- "Eliciting the help of a single expert to check the correctness of a mathematical proof."
  The system's main goal is to probe the mathematical abilities of candidate workers until it finds a competent worker for the task.

- "Getting game players to tag music clips in TagATune."
  In this case, players expect the system to provide a task, even if there are no more music clips to be annotated. In addition, players are likely to revisit the game if it is fun, e.g., the music clips should be neither too easy nor difficult to annotate, and chosen to both fit players' particular tastes in and knowledge about music.

- "Asking someone to take pictures of recycling boxes found near their current location."
  In participatory sensing, participants are chosen based on their proximity to the geographic location where data is needed [112]. Tasks are actively pushed to workers who are in the vicinity, who may or may not be willing or available to perform the task at that time. To be effective, a system needs to model the states of the workers (e.g., their availability) in order to decide whether or not to interrupt the users from their current activities.

4.1 PUSH VERSUS PULL APPROACHES

There are a wide variety of approaches for task routing. The actual choice of task routing method depends on many factors, including the following:

- **Workers' Expectation**. Is the human computation system expected to: (i) provide a task *only* if one is available; (ii) always provide a task (e.g., Duolingo, games with a purpose); or (iii) proactively seek out workers and assign them a task (e.g., in participatory sensing)?

- **Scale**. How many tasks are available? Tens? Hundreds? Millions? For example, if there are thousands of tasks we want workers to sort through, then having some way to visualize tasks would be helpful.

- **Nature of the Worker-System Interaction**. How long or how often are the workers expected to interact with the system? Can the system reliably infer the workers' characteristics? For example, if workers are transient (i.e., have sparse and intermittent interactions with the system), then the system might have difficulty modeling the expertise of workers; in this case, having workers select their own tasks may be more effective.

Conceptually, the two classes of task routing methods can be referred to as push versus pull. In the **push approach**, the system takes complete control over which tasks are assigned to whom, treats workers as passive receivers of tasks, and does not leverage human intelligence in the matching process. In games with a purpose, for example, this is the typical method for assigning tasks to players—i.e., players have no choice in deciding which objects (e.g., images, music or text) they are asked to annotate in each round of the game. Also, depending on whether the system has complete or partial information about tasks and workers, the task routing problem can be cast as an allocation/matching problem or an inference problem respectively. In contrast, in the **pull approach**, the system takes a passive role and merely sets up the right environment for workers to locate tasks themselves. The assumption is that human workers, when given the right tools for browsing, visualizing and searching for tasks, are equally or more capable than machines in assigning tasks to themselves or each other. For both push and pull methods, machine intelligence can play a significant role in helping to optimize the task routing process.

4.2   PUSH APPROACH

In this section, we will examine techniques for *pushing* tasks to workers automatically.

4.2.1   ALLOCATION

In some situations, the system assumes that it has full knowledge of the abilities of the workers as well as what abilities each task demands. For example, the system might know each worker's language skills (by pre-testing them ahead of time) and whether any given sentence requires a beginner, intermediate, or expert level translator. Here, the task routing problem can be reduced to the weighted exact set-cover problem for which an optimal solution is NP hard [288]. Another example is in participatory sensing, where each participant is associated with some costs and utilities (e.g., based on the geographic areas that he or she can cover), and the goal is to maximize coverage while keeping the costs within a certain budget. This problem is an instance of the budgeted maximum coverage problem, which is also known to be NP hard. In both cases, greedy algorithms [238] can be used to find an approximate solution.

When it is impossible to find a worker to perform a particular task, it is sometimes possible to break down that task into subtasks that workers can do, then assemble the results to generate the final output. For example, if the system cannot find a worker to translate a block of text from French to Italian, it can instead find two workers to translate the sentence first from French to English, then from English to Italian [288]. Making the assumption that the translations via each intermediate language incur some loss in quality, the task routing problem can be modeled as a multi-commodity flow problem, which can be solved using a linear program [288]. Again, the assumption here is that both task requirements and worker competence are known (or can be estimated) ahead of time.

4.2.2   MATCHING

The task routing problem is closely related to the "matching" problem [121, 134, 228, 274, 275], which has been studied extensively in the context of matching medical residents with hospitals, students to schools, etc. The matching problem, and more restrictively the "marriage problem," assumes that there are two sets of agents, where each agent has specified a complete set of transitive strict preferences over the other set of agents. The goal of matching is to finding a *stable* matching, where no agent prefers to be matched with anyone other than his current partner, by designing mechanisms that incentivize agents to reveal their true preferences. Although the connections are not yet established, it is possible that many of the same techniques and protocols created for the matching problem can be applied to match workers with tasks.

4.2.3   INFERENCE

The techniques described so far assume that the outcomes of assigning any task to any worker are known ahead of time. This assumption of perfect knowledge allows the task routing problem to be posed as a optimization problem and solved using mathematical programming. In reality, the system faces many uncertainties about both tasks and workers. The system can attempt to classify a task automatically into a domain (e.g., classify the topic of a search query) to be matched to the expertise of the workers, but often the classification accuracy is low (in fact, this is why human computation is needed in the first place). In some applications, pre-testing workers is not really possible or desirable. In other words, human computation systems are often faced with the situation that workers' expertise, motivation, and costs are unknown. In the presence of uncertainty, the task router must decide whether to spend time gauging the abilities of the unknown workers, or assign tasks to workers who are believed to be the current best candidates for that task. In this section, we will review some of these online, decision-theoretic algorithms for task routing.

4.2.3.1   Decision-Theoretic Models

In a decision-theoretic problem [206, 241], we assume that there are a set of actions, each of which is associated with a probability distribution over possible outcomes given the current state. In addition, each outcome is associated with some *utility* value that indicates its worth to the decision maker. The goal of decision making is to select actions with the highest *expected utility*. In the case of task routing, an action refers to the choice of a worker to perform a particular task, and the outcome utility of that action can depend on a combination of factors, including the quality of the task output and the worker's level of satisfaction (e.g., whether the experience of performing the task was enjoyable or fulfilling).

There are a variety of ways to model the utility of a task routing decision, as shown in the work on *proactive learning* [92]. Proactive learning is a variation of the active learning [284] framework that assumes the presence of multiple imperfect oracles. The goal of the proactive learner is to train a classifier using a fixed budget, by making as few queries as possible to the "best" oracles. Each task routing decision δ can be represented by the risk-utility tuple $\{R(\delta), U(\delta)\}$, where the risk $R(\delta)$ is the cost of the worker, and $U(\delta)$ is the utility of having the task done by the worker, which depends on the worker's characteristics.

Using this formulation, different types of workers can be modeled. For example, assigning a task t to a *reliable* worker has utility $U(t)$, but the same task assigned to a *reluctant* worker w (who sometimes refuses to answer) would have utility $P(\text{answer}|t, w) \cdot V(t)$, where $P(\text{answer}|t, w)$ is the probability that the worker will answer the query in the first place. The work of Donmez et al. [92] simulated additional types of workers, including fallible and infallible workers, and workers with cost that is uniform versus varying across tasks. Their algorithm has a *discovery* phase for probing the characteristics of each worker (e.g., the probability that a reluctant worker will answer a query), and a *task assignment* phase, where task-worker pairs are iteratively chosen to maximize the cost-benefit tradeoff. Different from other decision-theoretic approaches for data acquisition [223, 224], proactive learning assumes the existence of multiple, imperfect oracles and incorporates their distinct utilities when calculating the cost-benefit tradeoff.

Decision-theoretic solutions have so far only been used to answer the question of to whom a task should be assigned. Other dimensions of task routing, such as time, are left largely unexplored. One might expect human computation systems to be able to not only choose the best worker for a task, but assign the task at the best time and location for that worker. For example, the system can assign restaurant rating tasks to workers only when they are on location at the restaurant [120]. There has been also a lot of work on mixed-initiative systems, e.g., BusyBody [148], that infer the best time to probe users for information [169, 170]. Although not directly applied to human computation at the moment, these mixed-initiative techniques may be powerful additions to future human computation systems, enabling them to reason about time and space, and actively pushing tasks to workers (e.g., via mobile devices [98, 112]) without being overly disruptive.

4.2.3.2 Exploration-Exploitation Tradeoff

Instead of the two phase procedure (i.e., used in [92]) of first estimating the utility of each worker, then performing the task assignments to maximize the estimated utilities, there are algorithms that interweave the estimation and task routing process. At every time step, such an algorithm can decide whether to *explore*, i.e., assign an *information gathering* [241] task to a worker in order to learn about his or her characteristics, or to *exploit*, i.e., assign a task to the worker that the system currently believes is the best for the task.

An example of an online algorithm that makes these exploration-exploitation tradeoffs is IEThres [94]. The idea is to adapt the Interval Estimation (IE) algorithm for selecting oracles (each with different levels of competence) for a labeling task. Interval Estimation [167, 168] is a technique for choosing actions (e.g., the action of assigning a task to a particular worker) that balance the exploration and exploitation tradeoff. For each action a_i, the IE algorithm keeps track of the number of times n_i the action has been executed and the number of times w_i that the execution was successful. At each time step, the algorithm estimates the $(1 - \alpha)$ confidence interval of the success probability of each action, and chooses the one with the highest upper bound. The upper interval value can be large either due to a high sample mean of the success probability, or due to the uncertainties in the estimates. As more actions were performed, the intervals shrink and the algorithm is able to then select the best workers for any task. IEThres was also used in [95] to select oracles with time-varying accuracy—the idea being that the accuracy of human workers is likely to change over time, becoming less accurate as they are fatigued or get bored, or more accurate as they gain skills and knowledge as they perform more tasks. In [95], a sequential Bayesian model was used to estimate the accuracy of a worker at time t based on previous observations; based on these accuracy estimates, IEThres was then used to select the best human oracles for the task. This idea of using variance to indicate which worker needs to be *explored* is also used in the online EM approach proposed in [346].

4.3   PULL APPROACH

In contrast to the push approach, pull-based task routing methods do not assign tasks to workers explicitly; instead, the system plays a supporting role by providing the means for workers to find tasks to assign themselves. Previous work has shown that given a choice, workers tend to choose tasks for which they have expertise, interests and understanding [188]. Pull methods can be more effective than push methods, especially for platforms that have high worker turnover, making it difficult to infer the characteristics of workers who may have completed only a few tasks. A potential problem with pull methods is coverage and starved queues [22, 66]—it is found that completion time on Mechanical Turk follows a power law [342], with a large number of HITs finishing quickly but a small number of them taking a very long time. Better interfaces that allow workers to find tasks quickly can alleviate some of these problems.

4.3.1   SEARCH AND VISUALIZATION

In Mechanical Turk, workers are provided with a simple search functionality to search for tasks to do. This search functionality is very limiting—the work of Chilton et al. [64] found that workers end up constantly refreshing the page in order to find the most recently posted tasks to do. This inefficiency affects workers and requesters alike—tasks that are associated with very few hits take longer to be completed, because it is difficult for workers to find them in the first place. On other crowdsourcing platforms such as oDesk, tasks are organized by topics. There is no comparison yet of different search interfaces for task routing.

4.3.2   TASK RECOMMENDATION

Exemplified by services such as Amazon and Netflix, recommendation systems are a technology for matching users to a set of items (e.g., movie, product, music) that might interest them. The recommendation problem is generally discussed in terms of predicting missing ratings given a sparse matrix of known ratings of m items by n users. Content-based recommendation methods [19, 269] attempt to find similarities between user profiles and item characteristics. Alternatively, collaborative filtering methods [28, 199, 267, 278] make use of preference information about items (e.g., ratings) to infer the similarity of different individuals, and recommend items that users with similar tastes have previously consumed. Hybrid recommendation systems use a mix of content-based and collaborative filtering methods [48, 99, 198]. There has also been work on using statistical models to predict user preferences using item ratings as features [36].

Similarly, in human computation systems, one can imagine using similar techniques to recommend tasks to workers while resisting attacks [49, 187], as well as evaluate the success of the task recommendation system [137]. Already, recommendation algorithms have been used in online communities for helping users find tasks for which they have expertise, e.g., in Wikipedia [74] and Question and Answer forums [363, 364]. What might be very different in the case of human computation is that the objectives of the task recommendation system must be more than just satisfying the individuals, but also the system's requirements to generate accurate output. Another complication is that unlike products, tasks are much less persistent, i.e., they expire or disappear from the pool when completed.

4.3.3   PEER ROUTING

Instead of having the task router assign tasks to workers, or workers assigning themselves tasks, one can imagine a task routing system powered entirely by workers routing tasks to one another [361]. We call such a task routing approach *peer routing*. Conceptually, the traditional way in which academic papers are assigned to reviewers can be considered a peer routing system, since papers are passed from person to person until someone agrees that the paper is appropriate for their expertise.

Consider a peer routing system in which workers are given three choices—they can either accept a task, reject a task, or recommend another worker who may have more expertise to handle the task. In this case, people no longer need to visualize tasks; instead, they need to search for [210] and visualize other workers' expertise. A topic closely related to this is called *expertise location and recommendation* [225], which studies how technology can support the way people look for expert to help solve problems, and more broadly the problem of social matching [313], where the goal is to match people to one another using various forms of social data, e.g., profiles.

4.4   EVALUATION CRITERIA

In the previous section, we have described various push versus pull approaches for task routing. How do we compare the relative merits or different task routing policies? In human computation systems, the decision of what task to route to whom must take into account the objectives of both the system and the workers. From the system's point of view, how good a worker-to-task matching is depends partly on deadlines and priorities. Does the task need to be done right away by whoever that comes along? Is it critical that the task be done perfectly, i.e., by an expert? From a worker's point of view, the tasks that are assigned to them should reinforce their motivation in using the system, whether it be monetary reward or other intrinsic incentives.

In short, to evaluate a task routing policy, one can ask whether it leads to more accurate outputs (**accuracy**), tasks being discovered faster by workers (**discovery**) and completed more quickly (**efficiency**), and workers being satisfied (**worker motivation**). This is by no means a final and exhaustive list of criteria, but offers a starting point for discussing the merits of different task routing methods.

4.5   SUMMARY

This chapter addresses the task routing problem—how to match workers with tasks to satisfy both the system and the workers. The key points of this chapter include:

- Task routing methods can be divided into push versus pull approaches.

- Push methods can be cast as an allocation/matching versus an inference problem, depending on whether the system can assume complete or only partial knowledge of workers and tasks.

- Pull methods leverage human intelligence for routing tasks; the system simply provides workers with the means—e.g., interfaces to browse, visualize and search for tasks and other workers' expertise—to find the best matchings.

PART II

Design

CHAPTER 5

Understanding Workers and Requesters

In human computation systems, a *market* refers to a pool of individuals who are available to work on the computation tasks at hand. The possible motivations for human computers to want to participate are varied, but there is one thing in common: the particular computational problem that workers decide to devote time and effort to help solve has significant value to them. This value is often more than just monetary. Workers might be seeking access to valuable resources, entertainment, the opportunity to contribute to the common good or learn something new; in return, they are willing to perform small units of computation. On the other hand, requesters are also the stakeholders of any human computation systems, whose goal is to solve the computational problem of interests in the most accurate, efficient and economical way possible. A human computation system is not sustainable without satisfying the needs and wants of both workers and requesters.

Usability [239] is a concern in the design of any software systems; for human computation systems, it is no exception. First, a design that is easy to learn implies a low barrier of entry for new users and encourages them to revisit the system. This is especially important for human computation systems, such as games, where workers are volunteering their time and effort. If they do not find the task worth their while, it is not likely that they will continue to participate. Second, functionalities that are intuitive and personalized to each user allow tasks to be retrieved and accomplished more quickly. For example, it was found that the completion time for tasks approximates a power law [156]: many tasks get completed quickly and a few tasks are left incomplete and eventually expire. Without providing better search interfaces [64], this inefficiency is likely to persist. Finally, human computation systems present a set of unique design questions. To name a few: should workers/requesters be shown information about other workers/requesters (e.g., productivity level, trustworthiness) in order to induce competition within the market? What kinds of social norms and forces (e.g., reputation) encourage truthful behavior [186] and how can the system provide functionalities to support them? Can we build interfaces that allow users to collaborate on tasks? How do we design a game website that both fosters a sense of competition and community? How do these design decisions affect the quality and quantity of work done?

The questions are many; but the answers are few, as research on user-centric design of human computation systems is still in its early stages. The key point is that **understanding workers and requesters** and **designing for their needs and wants** is indispensable to a successful human computation system. In order to draw in and retain participation, human computation systems must be designed in such a way that they simultaneously provide the value that the workers and requesters seek while performing the intended computation. In this chapter, we will describe several markets and the characteristics of the workers—who they are and what motivates their participation. Finally, we will highlight the kinds of technologies that can be used to support workers and requesters as end users, and suggest new technologies that might fill the gaps.

5.1   MARKETS

5.1.1   MECHANICAL TURK AND PAID CROWDSOURCING

There exist many different crowdsourcing platforms, a list of which can be found at [152]. The most well known is Amazon Mechanical Turk (AMT), released in late 2005 [6]. The name Mechanical Turk is borrowed from a 18th century machine called "The Turk," a seemingly automatic chess-playing machine that is actually operated by a human in the background. Amazon Mechanical Turk provides a platform for *requesters* to post tasks to *workers* to perform in return for monetary payment. These tasks are called HITS, which stands for "Human Intelligence Tasks." This service quickly grew in popularity. By 2010, there has been an estimated 400,000 workers on Mechanical Turk [272].

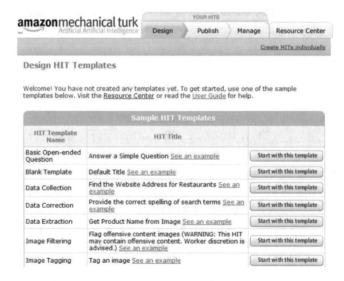

Figure 5.1: Requester interface.

Tasks distributed through Mechanical Turk are typically small (i.e., quick to complete), as are the monetary reward for each task—90% of the HITs have a reward of less than 10 cents [156]. Typical tasks include classification (e.g., images, music, documents), transcription, as well as the creation of original content (reviews, stories, blog posts) [156]. Psychologists, sociologists and economics are beginning to distribute their experiments, previously done in a laboratory, as tasks on Mechanical Turk [102, 146, 219, 246, 246, 307], because of the lower cost and comparable results [246] as well as the access to a larger, more global and heterogeneous pool of subjects [219]. Tasks can be created programmatically through an API, or manually based on a set of templates provided by the Mechanical Turk Requester Interface [7], as shown in Figure 5.1.

Anyone can sign up to become a worker on Mechanical Turk. For workers, Mechanical Turk provides functionalities (see Figure 5.2) for searching for tasks by keywords and minimum payment, and allows workers to order results by the recency, number of assignments, reward amount, expiration date, title and duration of the HITs. Upon the completion of a task and the approval of the requester, workers earn the predefined reward amount, sometimes supplemented with a bonus.

Figure 5.2: Worker interface.

There has been research documenting the demographics of workers on Mechanical Turk using surveys [157, 219, 272]. In a survey conducted in 2010 of 1,000 Mechanical Turk workers [157], workers are found to represent 66 countries, with the majority ($\sim$ 80%) from the United States and India. The survey asked workers questions about their age, income and education level, marital status, household size, and their experiences on Mechanical Turk such as time spent per week, number of HITs completed, amount of money earned, and their primary motivation for working as Turkers.

One interesting conclusion from the survey [157] is that the characteristics of the workers and their motivation depend heavily on workers' cultural background. It was found that there are significantly more (> 2:1 ratio) female than male workers in the United States, while the reverse is true in India. Turkers are on average younger and have lower income than the general population. However, a much higher skew is observed amongst workers in India than the United States. Likewise, the motivations for doing tasks on Mechanical Turk are different for American and Indian workers. Workers from the United States reported their work on Mechanical Turk as a secondary source of income and as a source of entertainment; while workers from India see work on Mechanical Turk as a primary source of income. Similar findings were reported in [272] which, additionally, analyzes the shifts in the demographics of Turkers, using six surveys conducted at semi-regular intervals within a 20 month period. They reported that Turkers seem to be getting younger, have lower income but

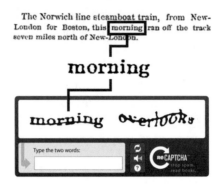

Figure 5.3: ReCAPTCHA.

higher education. Turkers report to earn, on average, only approximately $2.00 per hour, and some work for as many as 15 hours a week, making the work on Mechanical Turk a part-time job for some.

5.1.2   SECURITY AND ACCESS

CAPTCHAs [333] are widespread security measures used on the World Wide Web. You've seen them: images of squiggly characters on the Web that users must type to obtain free email accounts or buy tickets for a concert. Because computers cannot decipher the distorted characters, typing a CAPTCHA proves that you are a human and not a computer program that was written to abuse a service. For example, CAPTCHAs prevent ticket scalpers from writing programs to buy millions of tickets online. What most people don't realize is that by typing a CAPTCHA they are also participating in the largest distributed human collaboration project in history.

It is estimated that over 200 million CAPTCHAs are typed every day, each taking roughly 10 seconds of human effort—that is equivalent to 500,000 hours per day. The reCAPTCHA project channels this effort into a dual purpose: transcribing books. Here's how. Physical books and other texts written before the computer age are currently being digitized en masse (e.g., by Google Books and the Internet Archive) to preserve human knowledge and make information more accessible. The pages are photographically scanned and the resulting bitmap images are transformed into text files using optical character recognition (OCR) software. This transformation into text enables the books to be indexed and searched. Unfortunately, OCR is far from perfect. In older prints where the ink has faded, OCR cannot recognize about 30% of the words. On the other hand, humans are far more accurate at transcribing such print.

With reCAPTCHA old print material is being transcribed, one word at a time, by people typing CAPTCHAs on the Internet. Whereas the original CAPTCHAs displayed images of random characters rendered by a computer, reCAPTCHA displays words taken from scanned texts that

OCR could not decipher. The solutions entered by humans are then used to improve the digitization process. However, to meet the goal of a CAPTCHA (differentiating humans from computers) the system must be able to verify the user's answer. To do this, reCAPTCHA gives the user two words, one for which the answer is not known and a second "control" word for which the answer is known. If the user correctly types the control word, the system assumes they are human and gains confidence that they also typed the unknown word correctly. To date, over 750 million people—more than 10% of humanity—have helped digitize at least one word through reCAPTCHA.

To be successful, CAPTCHAs (and reCAPTCHA) must be both **secure** and **usable**. There are challenges on both fronts. On the one hand, several works have challenged the security of simple CAPTCHAs by attempting to break them using automated algorithms [233, 310]. On the other hand, a perfectly secure CAPTCHA can hinder access if it cannot be solved within a reasonable amount of time by the user. The design of CAPTCHAs must hit the sweet spot finding the most secure distortion that is still solvable by a human [60, 356]. Sometimes, fixing usability issues can dramatically improve the success rate of users' ability to solve CAPTCHAs. As an example, consider the Audio CAPTCHA, where words are not displayed visually but read aloud, with obfuscation by background noise such as music or difficult-to-identify chatters. It was found that Audio CAPTCHAs are notoriously difficult to solve [34, 279], compared to its visual counterpart. In two separate studies, blind participants were only able to solve 46% [279] and 43% [34] of the Audio CAPTCHAs, while sighted participants can solve 80% to 90% of the visual CAPTCHAs. However, by optimizing the user interface (without changing the underlying audio CAPTCHAs), the success rate of blind users solving audio CAPTCHAs on the first try jumps to 68.5% [34]. The lesson here is important: an effective design of the computation-eliciting interfaces can directly impact the efficiency and accuracy of computation. Consult [234, 333, 339] for an in-depth review of research on CAPTCHAs.

5.1.3 GAMERS

It is estimated that over 200 million users, i.e., over 25% of all Internet users, play online games every week [154]. Many of the online gamers play a genre of games called *casual games*. Causal games, which have been coined "video games for the mass consumers," have several properties [154] that make them appealing to even those who do not normally consider themselves as gamers. First, casual games have low barrier to entry, e.g., they can be easily accessed online with minimum to no setup. Second, They typically have only a few simple controls, and therefore extremely easy to learn. Third, they are *non-punishing*, e.g., allowing players ample of opportunities to score. Fourth, they can be consumed within short periods of time, e.g., 5–20 minutes during work breaks. Finally, casual games are typically inclusive, gender-neutral, and contain little violent content. This makes casual games suitable for players of different ages and from all walks of life.

The idea of using causal games as a medium for computation was introduced by von Ahn and Dabbish in 2002, with the introduction of the ESP Game [334]. The idea is to engage pairs of players in a simple game, where they tag images independently and are rewarded when their tags

Figure 5.4: GWAP.com.

agree. Despite its simplicity, the ESP game was played by hundreds of thousands of people, rapidly generating keywords that can be used to index images to power Web image search. The ESP Game was licensed by Google in 2006, who made its own version called the Google Image Labeler [130]. Since then, there have been many human computation games developed to tackle a variety of AI problems. For example, GWAP.com (see Figure 5.4) now hosts, in addition to the ESP Game, six human computation games—TagATune (for collecting music tags), Verbosity (for collecting common sense facts), Squigl (for locating objects in images via tracing), Matchin (for ranking images by preferences), FlipIt (for measuring image similarity) and PopVideo (for collecting tags of videos). In Chapter 6, we will explain the mechanics behind these, and other, human computation games.

According to the 2008–2009 Causal Games White Paper released by The International Game Developers Association (IGDA) [154], gamers play causal games in order to "relax, pass time, socialize, or achieve certain goals and challenges," and typically do not see themselves as being "gamers." In terms of demographics, it was found that 74% of causal gamers are female between the ages of 30 to 45 years old, playing mostly puzzle, word and card games. Currently, there exists no equivalent survey on the demographics of gamers who are attracted *specifically* to human computation games; this information would be invaluable—we can avoid the pitfall of assigning computational tasks to those who are not capable of doing them. For example, if the majority of the players visiting GWAP.com are young kids, then asking players to verify facts about world politics may be infeasible.

Human computation games must be **fun**; but this is more difficult to achieve than in typical causal games. The reason is that the design of a human computation game is constrained somewhat by the computational problem we are trying to solve. As designers, we must strike a balance between fun and practicality. Take for example, the challenge of building a game to collect labeled data for training a music classification algorithm. The ideal datasets used in training supervised learning algorithms should have a large and roughly equal number of examples for each class. Now consider two different versions of human computation games that collect tags for music. One solution, as prescribed in the Listen Game [325], is to randomly draw a small set of tags (e.g., "classical," "slow," "violin") from a *pre-defined* pool of 82 tags, and have game players choose the ones that describe the music clips. Another solution is to allow a pair of players complete freedom in entering any tags that come to mind. TagATune [190], for example, gives a pair of players two pieces of music, allow them exchange tags freely, and reward players when they can guess correctly whether the two pieces of music are the same or different. TagATune proved to be extremely fun—it has been played by tens of thousands of players, collecting over a million annotations. However, the data it collected was also extremely noisy [189]. There are synonyms (e.g., "calm" versus "smoothing," "violins" versus "strings"), spelling mistakes, communications between players (e.g., "yes," "how are you," "different"), and compound phrases (e.g., "cookie monster vocal") that are associated with very few examples. Learning from such open vocabulary data requires the development of new methods [189]. TagATune is a prime example of sacrificing clean data (which is the system's objective) in order to make the game more fun and attractive to players.

5.1.4   CITIZEN SCIENCE

Science is data intensive. In order to test hypotheses about our natural environment, e.g., about climate patterns, species distribution and trajectories of stars, often we need to collect and analyze data over large geographical regions and many time periods. When carried out by a few scientists, this process is tedious, time-consuming and sometimes impossible. The idea of citizen science is to engage non-scientists in the collection and interpretation of data in order to answer some scientific questions. Citizen science projects existed as early as 1900s where people volunteered to collect daily climate data at regular time intervals and report their measurements via telegraph [131]. The National Audubon Society holds an annual Christmas Bird Count, which has lasted more than 100 years and engaged more than 50,000 birding enthusiasts in collecting bird count information in 2005 [23]. The American Association of Variable Star Observers (AAVSO) [353] has engaged amateur astronomers for more than a century in tracking the variation of brightness in stars.

With new Internet and mobile technologies at the fingertips, it is becoming easy for vast number of people all over the world to participate as citizen scientists, to collect field data to help answer specific scientific hypothesis, to map and monitor animal and plant species, and to annotate massive amount of scientifically interesting images and field recordings; all these are still beyond the capabilities of machines. Galaxy Zoo [200, 201], for example, has more than 200,000 participants from 113 countries making more than 100 million classifications of galaxies [261], resulting in new discoveries [51] and an expanded project called Zooinverse [365]. E-bird [97], an Internet-based citizen science project run by Cornell Lab of Ornithology, has over a period of five years attracted over 500,000 users and collected 21 million bird records [306]. For example, following the Deepwater Horizon disaster in 2010, E-bird organized volunteers to monitor birds and their damage along the coast (see Figure 5.5) in order to measure the impact of the oil spill and better target the restoration efforts. The Great SunFlower project involves over 77,00 people in planting sunflowers,

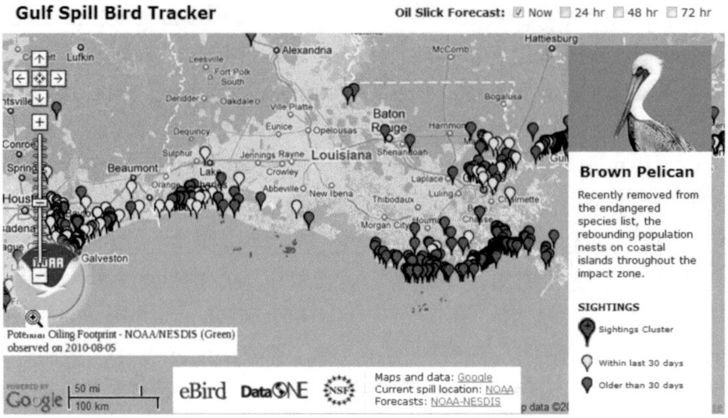

Figure 5.5: eBird sightings of birds during the Gulf Spill.

and observing the pollination activities of bees for 15 minutes daily [315]. Other projects involve citizen scientists in monitoring coral reefs [105], measuring precipitation [68], folding proteins [70], studying the evolution of species in response to climate change [104, 294], and genomic research [39]. New mobile devices and software (e.g., CyberTracker [76]) are now being deployed to enable people to collect information about their environment [208, 253], often in remote regions of the world [76]. A list of current citizen science projects can be founded using the ProjectFinder functionality in [281].

The goal of citizen science is to advance the state of science through the *collaboration* of citizen and professional scientists. Citizen scientists are driven by deep interests in the subject (e.g., birding, astronomy), passions about the cause (e.g., conservation and environmentalism) and a desire to contribute to research, but are often limited by the lack of knowledge and know-how. Professional scientists, on the other hand, are interested in collecting large amounts of data from citizen scientists towards answering scientific hypothesis or monitoring the environment; but they are also faced with the challenge of designing tasks that are easily understood and executable by citizen scientists, as well as detecting errors and biases in the crowdsourced data.

Much of human computation research can help address the goals and challenges faced by the two stakeholders in citizen science. First, technologies can be used to bridge the knowledge gap of citizen scientists. For example, human-in-the-loop computer vision algorithms can help novice birders identify birds in the field [43], and more intelligent task routing algorithms can help assign tasks to citizen scientists that fit their particular levels of expertise, interests and learning objectives. Second, user interfaces can be designed specifically to minimize data collection biases, e.g., over-estimation of more easily detected birds [306] or tendencies to over-report or under-report measurements of qualities (e.g., pollination level) due to its good (e.g., health of the ecosystem) or bad (e.g., existence of allergens) implications [253]. Likewise, professional scientists can benefit from using algorithms to distinguish between expert versus novice annotations, and filter out potentially erroneous data.

The most successful citizen science projects to date are ones that managed to turn a hobby (e.g., stargazing, birding, geocaching) into science, e.g., EBird and Zooinverse. Unfortunately, the reverse scenario is actually more common—often a scientist has some hypothesis that he or she wants answered, and attempts to transform a science into a hobby. Transforming a hobby into science is straight-forward because there is already buy-in from the participants, as well as longstanding best practices and culture that support the group and its activities. On the other hand, it is much more difficult to transform science into a hobby that tens of thousands of people would be interested in. In this case, turning data collection or analysis tasks into games may be an excellent solution.

5.1.5   LEARNERS

Problems that are hard for computers can also be transformed into tasks that are educational, so that students solve the problem at the same time as they learn.

The crux of this approach is "learning by doing." To illustrate the concept, notice that people could learn vocabulary in a foreign language with a system that repeatedly asks them to tag images from the Web. A beginner student who does not know how to say "dog" in Spanish may not be able to tag the images of dogs. However, if sometimes she is presented images for which tags are already known and she is told the correct answer when she cannot tag them, she may eventually learn that images of dogs should be tagged with "perro." Furthermore, this learner's performance, e.g., her success rate at labeling dog images, can be evaluated. Suppose now that this beginner student is presented with an image with no known tags, her answers could be used as potential tags for these images, weighted by her past performance. For instance, if recently she always tags known images of dogs correctly, the system would assign some confidence to her saying that an unknown image has a dog. Once multiple accurate students enter the same tag for a new image, the system would be confident that the tag is correct. In this manner, image tagging and vocabulary learning can be combined into a single activity.

Duolingo is a system built around this idea, which (among other things) aims to encourage millions of people to translate Web content for free into multiple languages. While human translators could be paid to do this task on Mechanical Turk, translating even a small fraction of the Web would be prohibitively expensive. For example, the Spanish version of Wikipedia has less than 20% of the content found in the English version; translating the rest of the English content into Spanish would cost approximately $50,000,000, assuming the low rate of 5 cents per word. While automated machine translation is another avenue, computers cannot yet translate language nearly as well as humans can. Although one goal of Duolingo is translation, students in this language-learning site accomplish many other human computation tasks as well. While text translation can be used to teach grammar and vocabulary, it cannot be used to teach speaking or listening skills. However, such skills can be taught by asking students to perform other valuable tasks that are hard for computers, such as transcribing audio clips or subtitling videos. From the user's perspective, Duolingo is a complete and free language learning service; from the human computation perspective, the site's purpose is to collect valuable work from the students.

Learners exist not only in the domain of language learning. The Cornell Lab or Ornithology citizen science project [72], for example, make it their special mission to "help participants learn about birds and experience the process by which scientific investigations are conducted." [40, 319] It was found that participants were able to improve on their knowledge about bird biology as well as gain a better understanding of the scientific processes [45, 318]. In addition, citizen scientists usually have deep interests in the subject area and the desire to learn. In a study examining the motivation of Galaxy Zoo volunteers, it was found that "interests in astronomy" and the sense of "wonder about the vastness of the universe" are what motivate volunteers to participate [261].

A human computation system that is capable of teaching a skill, whether it be learning a language or distinguishing birds, creates a win-win situation: it can help volunteers become better contributors, which in turn improves the efficiency and accuracy of the computation. The research questions are plenty; to name a few—How do we build algorithms that can automatically select the appropriate lessons for each learner? How does the system decide how long to teach learners, and at what point their answers can be trusted? What kind of user interfaces are conducive to learning? Can we identify expert workers and pair them up with novices, so that they can help teach skills that the system has not yet mastered? The research into the answers for these questions is in its infancy. Nevertheless, the idea that expertise can be *cultivated*, we argue, is an important driving force for next-generation crowdsourcing applications, as there are typically many more non-experts than experts in the world. Studying effective ways of creating, and not simply gathering, a crowd of experts is an important future direction for human computation.

5.1.6   TEMPORARY MARKETS

Some workers are motivated to participate by urgent but temporary circumstances, e.g., natural disaster [236, 301] such as earthquakes, or concerns about water quality in the neighborhood. These workers can typically be engaged quickly when the event happens, but after the event, it is hard to retain the workers and have them continue performing the tasks, or to guarantee that the same workers will return during a similar event that happens in the future. Many open questions remain as to how to design human computation systems to recruit and retain workers from these temporary markets.

5.2   SUPPORTING END USERS

Workers and requesters can be considered the two main end users of human computation systems. In this section, we will highlight ways in which technology can be used to support their objectives and sustain their participation.

5.2.1   WORKERS

Workers, even those belonging to the same market, are often motivated by a diverse set of objectives. However, regardless of the particular objectives, the system can support workers by providing them with the means to make progress, monitor progress, and handle obstacles that hinder progress towards achieving their objectives. These include functionalities that enable workers to:

- search for the *best* tasks, e.g., highest reward-to-effort ratio, most interesting, most educational, most relevant to one's expertise, quickest to complete, most similar to other previously liked tasks;

- make decisions about *which task* to perform and *when*, by visualizing the tradeoff between accepting one task over another, or now versus later;

- monitor progress and productivity (e.g., leader boards, task history, etc.);

- communicate problems and grievances about tasks, wages, and any other obstacles that hinder their progress;

- understand the competition, e.g., by visualizing the characteristics of other workers performing the same task (e.g., leader boards);

- find support within the community (e.g., chats, forum, etc.);

- express opinions about tasks and requesters (i.e., a reputation system).

A few of these points are predicated on the assumption that the online community is itself thriving. For this, there is substantial research on how to create, maintain and grow online communities [173, 184, 244, 258, 259].

5.2.2   REQUESTERS

Requesters are motivated by the need to solve some computational problem as accurately and efficiently as possible while minimizing cost. To support requesters, the system can provide requesters the means to:

- automate and optimize task design [150];

- visualize workers' expertise and interests (e.g., there is no need to even post a "Norwegian-to-Swedish" translation task if no worker is fluent in those languages);

- meet deadlines;

- meet budget constraints;

- understand the competition (e.g., by visualizing the tasks that other requesters have posted).

- express opinions about different workers.

Trends in recent research suggest that in the future, requesters will not need to interface directly with crowdsourcing platforms such as Mechanical Turk, but instead they will be accessing the crowd from everyday applications, such as the word processors [31] and databases [248]. Soylent [31] is a good example: by embedding a tool to access crowdsourcing services in a word processor, it enables common people to become requesters of human computation services. Another example is the work of Parameswaran et al. [248], which introduces a querying language that allows requesters to access Mechanical Turk as if it is a database where the facts are computed by humans. For example, the following complex task "Find photos that are small, in jpeg format, containing headshots of politicians with a plain background" can be represented as the following query:

travel(I) := rJpeg(I), hPolitician(I), haPlainBackground(I), aSmall(I), rDemocrat(I),

where the prefix h refers to a predicate to be evaluated by a human through a crowdsourcing service, and a refers to a predicate to be evaluated by an algorithm, and r refers to a predicate to be evaluated by looking up information in an extensional table (e.g., in this case, we assume that there is a table that contains a list of politicians and their political affiliations), and ha refers to a hybrid predicate that can be processed by a human or machine. The application designer must map each h-predicate to a task to be presented to a worker on a crowdsourcing platform. This language representation makes it easy to specify different operations to be performed by humans versus machines, and hide much of the details—e.g., performance/cost optimization, output aggregation, latency minimization, etc.—from the requesters. This view that Mechanical Turk is analogous to a database engine with humans in the loop is also conveyed by Qurk [217] and CrowdDB [115].

Finally, while we typically think of requesters as workers, in the future, requesters may be automated AI systems that request human feedback in order to self-monitor and improve. New technologies will be needed in order to enable this interface between AI and human computation systems.

5.3   SUMMARY

This chapter addresses the question of motivation, and highlights technology that can help address the needs and wants of the two main stakeholders of human computation systems: workers and requesters. The key points of this chapter include:

- There exist a variety of markets for human computation, including paid workers, CAPTCHA solvers, gamers, citizen scientists and students.

- Workers, even those that belong to the same market, often have heterogeneous motivations. A human computation system must be designed to provide multiple avenues for different workers to achieve their particular objectives.

- Increasingly, human computation systems will become black-box tools that will treat requesters as end users, by taking care of and hiding the complexity of the human computation process.

CHAPTER 6

The Art of Asking Questions

Querier: *What is the answer to 1+1?*
Machine: 2
Toddler: ?
Spammer: 364
Computer Scientist: 1
Philosopher: it depends ...

There are major differences between how humans and machines compute. For a simple question such as "what is the answer to 1+1?," one can safely assume that machines will give a consistent (i.e., always the same) answer. Unlike machines, the same question asked to *human computers* might yield different answers that are biased by their particular competence and expertise, intentions, interpretation of the question, personal preferences and opinions, and general physical and psychological limitations, e.g., fatigue, lack of motivation and cognitive overload. More importantly, humans can be *influenced* to perform more consistent and accurate computation. For example, a spammer [118] who is not rewarded unless his answer agrees with the majority is less likely to cheat. Likewise, a toddler who is asked to count oranges instead of abstract numbers, or a computer scientist who is told the correct interpretation of the question (i.e., that the "+" is the arithmetic, not logical, operation) can probably produce the correct answer.

Previously, we have reviewed two points of intervention for quality control: *before* computation, by routing tasks to the right worker (Chapter 4) and *after* computation, by aggregating and filtering outputs (Chapter 3). The hypothetical session above illustrates an important point: the way we ask questions to human computers can directly affect the output that they generate. This chapter describes safeguards placed *at* the time of computation, rather than before or after. In particular, we will discuss how the design of tasks and mechanisms can influence the way human computers compute—motivating them to tell the truth, enhancing (or degrading) the quality of their outputs, or making them reach an answer faster.

6.1   DESIGNING TASKS

By our definition, a **task** refers to an actual piece of work (e.g., Human Intelligence Task) that is performed by a human worker, which can map to an operation, control or synthesis process in the actual algorithm. Note that each task may encompass multiple processes. For example, it might be easier to get the same workers to generate an intent for a search query *and* judge the relevance of the webpage according to the intent, than to split the task between two different workers.

Typically, a task consists of three main components:

- **Basic Information.** This includes (a) inputs, (b) what is being computed (i.e., what is the question being asked about the inputs), and (c) allowable outputs. For example, for an image annotation task, the input is an image, the question is "what are some objects found in this image," and the allowable outputs are tags.

- **Conditions for Success.** How how will the computational task be considered successfully done? For example, in the ESP Game, the condition of success is that at least one of the outputs between the two players match.

- **Incentives.** How will the successful (or failed) execution of a computational task be rewarded (or penalized)? For example, in the ESP Game, players receive 10 points for matching, and 0 points otherwise.

Tasks need to be properly designed so that the human workers are actually computing the correct outputs that the algorithm needs. In this section, we will review five design decisions that go into creating a task, including

- **Information.** How does the information presented to workers influence their task performance? Do workers benefit from having seen other workers' solution? Are the instructions precise and unambiguous, i.e., will workers interpret the task in such a way that actually computes the intended function?

- **Granularity**. Is the task well defined or should it be decomposed into simpler subtasks? Is the task cognitively overwhelming for the user to do?

- **Independence.** Must the task be done independently by workers? Can workers communicate with each other and collaborate?

- **Incentives.** Is the payment adequate for the amount of efforts required by the task? Would workers be motivated to perform the task to the best of their abilities?

- **Quality Control.** How do we ensure that the outcome of the task is correct?

These design decisions have inter-dependencies that are intricate and little understood. For the rest of the section, we will discuss each of these design considerations in turn.

6.1.1   INFORMATION

There are plenty of psychology experiments which show that human subjects can be systematically biased by how a question is presented and what information is included. For example, it was found that people often estimate by "anchoring," i.e., starting with a reference value and adjusting from it. When asked to estimate the percentage of African nations who are members of the UN, people who were asked "Was it more or less than 10%?" gave lower estimates than if they were asked "Was it more or less than 65%?" [326]. Likewise, people can anchor to items that are presented earlier in a series, e.g., their estimates for "$1 \times 2 \times 3 \times 4 \times 5 \times 6 \times 7 \times 8$" were smaller than for "$8 \times 7 \times 6 \times 5 \times 4 \times 3 \times 2 \times 1$," even though the two computational problems are equivalent. How questions are phrased can affect how fast people can compute. For example, it was found that people can perform calculations faster and more accurately, if the question is formulated as a scenario that is familiar to them (e.g., farmers can calculate better if the calculations are about produce than abstract numbers).

The list of cognitive biases is long [250, 251, 326, 351], but there has been little research on the effects they have on the way workers perform computational tasks. For example, crowd workers often perform a sequence of assignments belonging to the same HIT. When workers perform tasks in sequence, they are subject to sequential context biases [80, 302], such as assimilation (i.e., the response to the current stimulus is larger when the previous stimulus is of greater intensity, and smaller when the previous stimulus is of lesser intensity) and contrast (i.e., the response to the current stimulus is smaller when the previous stimulus is of greater intensity, and larger when the previous stimulus is of lesser intensity). This implies that, for example, if the task is to rate the aesthetics of an image, workers' rating of one image can depend greatly on how beautiful or ugly the preceding image was. There has been some work that attempts to automatically remove sequential dependencies in a series of rating so that the ratings actually correspond to workers' true beliefs [235].

Information can sometimes help a worker perform a computational task faster or more accurately. For example, included with the task can be a **partial solution**, generated either by a machine or another human worker, that serves as a starting point for the worker to tackle the problem. In the FoldIt [69] game, it was found that when given a partially solved conformation, players were much more capable of arriving at a solution than if they started from a blank slate. Likewise, Little et al. [202, 203] found that for some tasks, it is beneficial to have workers improve (or *iterate*) on existing solutions. In an image description task, it was found that by having people iteratively improve upon previous descriptions, the final description of the image is on average longer and more descriptive. On the flip side, showing existing solutions can sometimes hurt. For the task of brainstorming (e.g., inventing a name for a fictitious company based on its description), iteration seems to hamper creativity, causing people to repeat words that were in previous solutions instead of thinking of new names. Other works have found that in iterative tasks, the earlier solutions set a strong example for future solutions [185, 193], an effect that has been referred to as *seeding* [175]. For example, in one study [193], workers are asked to iteratively provide an additional step (e.g., "book a plane ticket") towards achieving a high-level mission (e.g., "travel to France). When the first worker's answer is actually a short phrase, subsequent workers will mimic and also enter short phrases; likewise, when previous workers entered elaborate paragraphs, this also influences other workers to do the same. In other words, previous workers' outputs can have either positive or negative effects down the line.

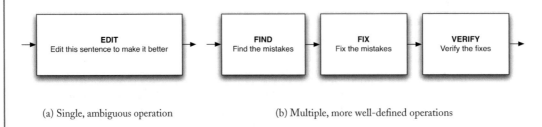

(a) Single, ambiguous operation   (b) Multiple, more well-defined operations

Figure 6.1: Task granularity.

6.1.2   GRANULARITY

Tasks that require multiple steps or have ambiguous instructions/requirements may be better off being decomposed into a set of shorter, more well-defined tasks. One example is crowdsourced text editing, where it was shown to be more effective to break down each editing task into a sequence of three subtasks [31], having some workers highlight the errors in the text ("find" operation), other workers fix the errors in the text ("fix" operation), and having yet other workers confirm that the corrections are valid ("verify" operation). This find-fix-verify pattern (Figure 6.1) yields better results because it helps to control the variance of efforts amongst workers—preventing the *lazy* worker from spending the minimal effort and picking the easiest errors to fix, and preventing the *eager* worker from over-correction.

The question of how to design a task with the appropriate effort requirement is sometimes best answered empirically, by testing different combination of parameters and how they affect the quality of the output [150]. For example, Huang et al. [150] ran 38 different versions of the image labeling task (that differ in terms of the number of images to label, the number of tags to collect per image, the total number of HITS, and the reward per HIT) and use the results as training data to predict the optimal task design to meet different objectives. For example, for the objective of maximizing the number of unique tags, the optimal design is to create more time-consuming, but higher paying HITS, in order to save on posting cost charged by Mechanical Turk.

6.1.3   INDEPENDENCE

In addition to simple tasks (e.g., image labeling), human computation can be used to handle complex tasks that require creativity, innovation and collaboration, e.g., subjective user studies [176], collaborative translation [175], collaborative authoring [56], etc. While it is common practice that workers perform tasks alone, there are now new platforms emerging that will allow workers to perform tasks by interacting with each other [175, 221]. Winter et al. [221] created a web application that allows Turkers to simultaneously perform a task (e.g., play an oil field exploration game), in order to study the otherwise hard-to-observe collaborative human behavior. In another experiment, workers are asked to collaboratively translate a poem using Etherpad [175], a collaborative editor.

It was found that by allowing workers to coordinate their edits and exchange ideas through chat, workers voluntarily returned to do more, even after having been paid for their work.

Collaborative tasks have interesting implications. Instead of identifying a single worker for a task, the task router must now find a group of workers with the right mix of expertise. Task routing, therefore, becomes the problem of **team formation**.

6.1.4 INCENTIVES

Another way to influence the output of human computers is through reward and punishment. Incentives may affect many aspects of workers' behavior, including whether they do any task at all (i.e., level of participation), which task they choose (i.e., the quality of task assignment), and how well they perform each task (i.e., the accuracy and efficiency of the computation). This, in turn, affects requesters; in crowdsourcing marketplaces where there are multiple requesters competing for the same human resources, prices need to be competitive [156] in order for tasks to be chosen by workers. There are a few questions to ask when designing an incentive system, including: (1) the form of incentive (e.g., extrinsic versus intrinsic) to use; (2) how much incentive to provide; and (3) ways to ensure that the incentive system is resistant to manipulation.

The particular form of incentives needs to align with what the workers are actually motivated by [10, 58, 280, 289]. Some workers have **extrinsic motivation**, e.g., monetary payment and virtual rewards (e.g., points, badges), while others have **intrinsic motivations** [266], e.g., the desire to influence (power), to know (curiosity), for social standing (status), for companionship and play (social contact), to get even (competition), to improve society (idealism) and to collect (ownership). The interaction between extrinsic and intrinsic motivation is complex [81, 117], e.g., paying for tasks that participants are intrinsically motivated to do might actually demotivate them. The usage of incentives to motivate participation is a central issue in the study of online communities; consult [184] for a comprehensive review.

While there is mixed evidence on whether greater incentive actually leads to better work [9, 220], there is general consensus that incentives are an important motivator for participation. The question of how much incentive to provide—e.g., how much to pay for each task, or how many points to assign to each round of game, is itself a complex one that requires some understanding about the psychology of the workers. For example, the so-called "target earner" are motivated to do tasks until they reach a target earning that they personally define [50, 107, 110, 145]; in which case, a pricing structure that allow them to clearly visualize their progress towards this target (e.g., pricing tasks at a value divisible by five [145]) might help motivate their participation. There has been evidence also that paying the right price is important [109, 128], i.e., sometimes not paying at all is actually better than paying too little.

Incentives should encourage the desired behaviors in workers, e.g., to not game the system, select tasks one has expertise in as opposed to just easy tasks that are quick to complete. There has been a lot of work on using incentives to engineer the right behavior in online communities, in the context of question and answering websites [124, 126, 160, 237], peer prediction systems [184, 229] and contest design [90, 231, 232, 320]. Much of the work in game theory and mechanism design (and more specifically, work in the team problem or combinatorial agency [5, 16, 138, 162]) are relevant for designing manipulation-resistant incentive systems in human computation.

Finally, most human computation systems use fixed pricing that is determined ahead of time by the requesters. In the future, task pricing may be variable (e.g., determined based on the quality of the outcome, such as proportional to how much better a solution is compared to previous solutions [69, 73]) and determined by the workers themselves (e.g., workers can sell their service at a bid price [146, 194]).

6.1.5   QUALITY CONTROL

Last but not least, one can design quality control tasks to check whether the output of a completed task is correct. Some of the most common quality control tasks include:

- **Verification.** Check if an output is correct [15, 305].

- **Voting.** Vote for the best output.

- **Filtering.** Vote for the worst output.

- **Merging.** Aggregate a set of outputs into a single solution (e.g., merging several outlines of an essay [178]).

Note that verification tasks also need redundancy, i.e., multiple workers should perform the same quality control tasks. For example, to select the best description for an image, multiple workers can be asked to perform the *voting* task; the best description will be the one that receives the most votes.

Beyond quality control tasks, several works mentioned the possibility of leveraging social forces and protocols—e.g., social norms and sanctions, monitoring, legal contracts, promise of future work based on current reputation [175] and sense of community [186]—to reduce gaming of the system. Kittur [175] describes two ways to design tasks to control for quality: (a) ensure that the amount of effort taken to produce rubbish answers is the same as that needed to produce good answers, and (b) signal to workers that their work is being monitored. For example, placing verifiable questions before subjective ones resulted in the percentage of invalid comments dropping by 43%, the doubling of effort, as well as high quality of work [176].

6.2   ELICITING TRUTHFUL RESPONSES

The objective of a human computation system to generate high-quality outputs is not achievable if the human computers are unmotivated or unwilling to tell the truth in the first place. One way to elicit truthful responses from workers is to design a set of rules for interacting with the system in which workers benefit the most by being truthful. This set of rules, referred to as a *mechanism* [159, 240], defines the set of permissible actions for the worker and specifies how the final outcome will be computed based on those actions.

Our use of the term *mechanism* is borrowed from a field of research called *mechanism design* [240], which studies systems in which multiple self-interested agents hold private information (that we want revealed truthfully) that is essential to the computation of a globally optimal solution. Because each participant is considered *rational*, i.e., with the selfish goal of maximizing one's own expected payoff, he or she may want to withhold or falsify information. In order to achieve the best economic outcomes, the goal of the system designer is to find a set of rules in which participants benefit the most by sharing their private information truthfully. A canonical example is the Vickrey-Clarke-Groves (VCG) mechanism—when used in a single-item auction, the mechanism specifies that the agent with the highest bid should receive the item, but charged the price of the second highest bid. It has been shown that the VCG mechanism is *incentive-compatible* [153]—the bidders cannot do better by mis-reporting their true valuation of the item. An important premise behind mechanism design is that agents cannot be instructed, taught or forced to behave in a certain way; however, by designing rules that align the motivation of the agents with the objectives of the system, we can encourage agents to report their private information truthfully.

In human computation, the term *mechanism* takes on a similar but slightly different meaning. Here, the private information that the human computers hold, and that our system wants to collect, is the *true* output to a computation task. We design mechanisms—a set of rules governing how the output of each human computer will jointly determine an outcome (i.e., reward or penalty)—in order to incentivize human computers to produce outputs in a truthful way. In this section, we will show how these mechanisms can be embedded in multi-player online games. Games are a particularly powerful vehicle for computation as they have the potential to reach a huge number of willing participants over the Web. Exactly how to design a game to make it attractive to players is a more difficult question, although some work have offered theories [213, 308] about what makes games fun.

6.2.1   HUMAN COMPUTATION GAMES

Human computation games, or the so-called *games with a purpose* [336], are multi-player online games where players generate useful data as a by-product of play. The first human computation game, the ESP Game [334] (Figure 6.2), was created to collect tags that can be used to describe and index images, making them easily retrievable on the Internet. In this game, two players are given the same image, and asked to independently enter tags that describe that image. Upon agreeing on a tag, the players are rewarded with points and the image is successfully labeled.

To motivate truthful responses, the ESP Game combines three ideas—independence, agreement and shared information. By having two players independently generate the same tag, the system has higher confidence that the tag is correct than if the tag is generated by a single person. Furthermore, the only common information that the two players share is the image; in the absence of extra information (i.e., assuming that players do not communicate with each other), players are

Figure 6.2: The ESP game.

more likely to find a matching tag if they limit themselves to only the tags that are relevant to the image, a search space that is much smaller than the set of all words in the English language. The ESP Game is a specific instance of a mechanism called "output-agreement," where two players get the same input and are rewarded when their outputs agree. Mechanisms are generalizable—in fact, the output-agreement mechanism has been successfully applied to other problems, including image preference [29, 133], music classification [215, 325], ontology construction [296, 328] and sentiment analysis [282].

6.2.2   LEVERAGING COMMUNICATION

Mechanisms are generalizable, but not one size fits all. The output-agreement mechanism relies on the fact that it is relatively easy to match on a tag for an image. When the ESP Game was adapted for audio tagging, it was found that players have much greater difficulty matching on a tag. For example, in a prototype game that uses the output-agreement mechanism to tag music, it was found that in 36% of the rounds, players opted to pass after a few tries, giving up on trying to match [191].

The input-agreement mechanism is a solution to this problem. In this mechanism, players are each given an input for which they have to generate some outputs. In addition, they are given an auxiliary function to compute that depends on *both* of their outputs. In order to successfully compute this auxiliary function and receive rewards, players are allowed to communicate with each other and exchange messages about their outputs. A game called TagATune (as depicted in Figure 6.3) implements this mechanism. In this game, players are each given a music clip and asked to guess whether the music clips are the same or different, after exchanging tags with each other. The mechanism behind TagATune leverages the fact that players must be truthful to one another in order

Figure 6.3: Tagatune.

to succeed in determining if the music clips are the same or different. The human computation system can then collect accurate descriptions of music by "eavesdropping" on the players' conversations.

This mechanism belongs to a class of mechanisms called *function computation* mechanisms, where players are given some partial input (e.g., a music clip) for which they need to perform some computation (e.g., generate tags), in order to compute an auxiliary function (e.g., whether the two pieces of music are the same or different) which determines the reward of both players. The TagATune mechanism is a specific instance of the function computation mechanism called *input-agreement* [335], where the function to compute is 1 if the input objects given to the two players are the same, 0 if they are different. Like the ESP Game, the input-agreement mechanism is general (i.e., can be applied to annotate other types of data, e.g., images, video, or text) and is particularly useful for input data that has high description entropy (i.e., can be described in many semantically equivalent ways) [190].

Verbosity [300] and Peekaboom [337] are two examples of human computation games that use an asymmetric version of the *function computation* mechanism to elicit truthful outputs from players. Verbosity [338] (Figure 6.4) is a game for collecting common sense facts. In this game, players alternate between the role of a *describer* and *guesser*. The describer is given a secret word (e.g., "crown") which he has describe to his partner, the guesser, by revealing clues about the secret word (e.g., "it is a kind of hat"). Both players are rewarded if the guesser is able to guess the secret word. The mechanism requires the guesser to compute the auxiliary function "what is the secret word" given the outputs (i.e., clues) of the describer. It is asymmetric in the sense that only one of the players is responsible for computing this auxiliary functional and that the communication is (mostly) unidirectional.

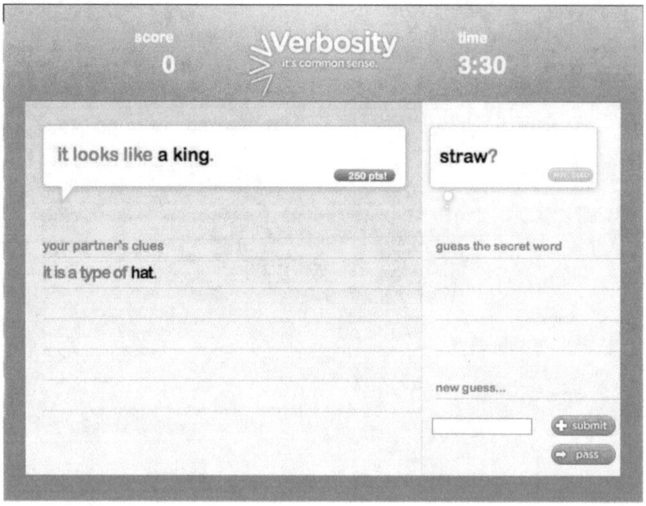

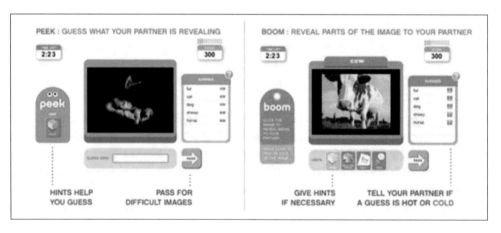

Figure 6.4: Verbosity.

Figure 6.5: Peekaboom.

Another example of an asymmetric function computation game is Peekaboom (Figure 6.5). Peekabom is a game for locating objects in images and involves two players—the *boomer* and the *peeker*. The boomer is given an image and a secret word (e.g., the word "cow) and must click on and reveal the part of the image associated with the secret word to his partner, the peeker. On the other hand, the peeker is initially given a blank image that is slowly unveiled by the boomer, and must guess the secret word as quickly as possible. Both players are rewarded when the peeker guesses the correct secret word.

In both Verbosity and Peekaboom, the auxiliary function asks "what is the secret input object your partner holds, given his or her descriptions of that object?" Asymmetric function computation mechanisms with this type of auxiliary functions are referred to as *inversion problem* mechanisms [335].

Designers of human computation games are faced with the challenge of building a system that simultaneously meet two (and often competing) objectives—to satisfy the players and to perform efficient and accurate computation. On the one hand, one can design a fun game that attracts a lot of players, but which does not collect any useful data. On the other hand, if the game is designed to collect the cleanest possible data, without paying attention to the enjoyability of the task, then no player would be interested. In function computation games, e.g., TagATune, Verbosity and Peekaboom, this tradeoff is apparent. Granting players more freedom of expression (e.g., allowing them to enter free form text, or communicate with each other) can make the game more entertaining, but can lead to noisy data that requires a great deal of post-processing. In Verbosity, a significant amount of filtering needs to be done to the collected data before they are considered trusted [300]. This is because the describers sometimes cheat by entering clues that are not common sense facts, but shortcuts for revealing the answer to the guesser [300]. Common cheats include clues about the number of letters in the secret word (e.g., "it has three letters"), or mnemonics (e.g., "it sounds like king"). Likewise, by allowing open communication between partners, the tags collected by TagATune are noisy (e.g., contain phrases such as "hi, how are you") and require substantial post-processing [189].

6.2.3   EXPLICITLY PREVENTING BAD OUTPUTS

A different way to elicit truthful outputs is to explicitly prevent players from entering outputs that the system considers undesirable (e.g., inaccurate or uninformative). A straightforward way would be to tell players *not* to enter particular kinds of outputs. However, as mentioned before, the premise behind mechanism design is that players are self-interested agents who are not guaranteed to follow instructions. A better way is to design a mechanism in which players are naturally motivated to avoid generating bad outputs.

Take for example the ESP Game. It has been noted that the image tags produced by the ESP Game tend to be common and uninformative [161, 343] . This is a direct consequence of the output-agreement mechanism—needing to agree with his partner, a player's best strategy is to enter common tags that are likely to be entered by any person. A possible remedy is to use rewards to motivate players to enter more specific tags—e.g., the Google Image Labeler gives players higher scores for more specific labels. Another solution is to impose restrictions on what the players are allowed to enter. In the ESP Game, *taboo words* [334] are introduced to prevent players from re-entering high frequency tags. None of these approaches seem to solve the problem completely [343]. In fact, in many output-agreement games, the improper use of restrictions can lead to bad results. For example, in the effort to collect a diverse set of results, the game Categorilla [329] forces players to enter only an answer that begins with a particular letter, without knowing if such an answer actually exists. As a result, players often have great difficulty coming up with such word, and end up

(a) "Positive Player"   (b) "Negative Player"

Figure 6.6: Polarity.

generating nonsensical answers instead. Likewise, in the prototype version of TagATune, players are prompted to enter tags belonging to a particular category (e.g., "instrumentation," "mood," "genre") chosen by the system at random without regards to whether such a category is actually relevant to the audio clip. For example, players are sometimes faced with the impossible task of describing the "mood" of an audio clip containing speech.

An alternative mechanism for generating more specific image tags was introduced in a game called KissKissBan [141]. This mechanism adds a third player (called the *blocker*) to the ESP Game who acts as an adversary, entering tags to block the other two players from matching. As a result, the players are constrained to enter more specific tags that the blocker is not likely to think of. Take as another example the task of fact verification, e.g., determining if a tag describes an image or not, or checking whether the named entity "Washington" is a "city," "politician," or "state." The output-agreement mechanism, in this case, would work poorly—if players are rewarded for agreeing, then there is a simple cheating strategy for players to state a fact to be true each time and receive the maximum reward. One way to prevent players from cheating is to inject facts that the system knows to be true or false in order to determine whether a player is gaming the system [136]. However, this necessitates a large collection of verified facts, which may not be easy to obtain. What we desire is a mechanism in which players are motivated to state a fact to be true if and only if it is actually true.

Polarity (Figure 6.6) is a fact verification game that implements the *complementary-agreement* mechanism. In this game, players are presented with a set of candidate facts to evaluate. For example, players are given the named entity (e.g., "Carnivores") and asked which tags are accurate descriptions of that named entity. Players alternate between two roles: the "positive" player (Figure 6.6(a)) is asked to select tags that describe the named entity, while the "negative player" (Figure 6.6(b)) is asked to select tags that *do not* describe the named entity. Players receive points for selecting as many tags as possible (more specifically, their joint score is the product of the number of tags each player selected), but if any of their selections overlap, they receive a penalty. Essentially, players are generating *complementary* data that is used to constrain each other's outputs. The positive player is motivated to select only facts that are true, in order to minimize the chances of an overlap with the negative player's outputs, vice versa.

6.2.4   A BRIEF SURVEY OF GAMES AND MECHANISMS

There has been a large number of human computation games invented to tackle different problems, many of them are identical in terms of the underlying mechanisms. Table 6.1 provides a summary of some existing human computation games and their underlying mechanisms; consult [314] for a comprehensive survey of human computation games for knowledge acquisition.

Table 6.1: Survey of human computation games.

Game	Description	Mechanism	AI Problem
The ESP Game [334]	two players match on a tag for the same image	output-agreement	object recognition
Peekaboom [337]	player 1 reveals parts of the image associated with a secret word, player 2 must guess the secret word	function computation (problem inversion)	object location
Verbosity [338]	player 1 describes the properties of the entity associated with a secret word, player 2 must guess the secret word	function computation (problem inversion)	knowledge extraction
TagATune [190]	two players exchange tags and determine if two music clips are the same or different	function computation (input agreement)	music classification
FoldIt [69]	players fold protein structures to minimize total energy	function computation (optimization)	protein folding
HerdIt [24]	players select tags that describe the music	output-agreement	music classification
Categorilla [329]	players name an entity that fits a template (e.g., Things that fly)	output-agreement	natural language processing
MoodSwings [174]	players click on a 2-dimensional grid to indicate the valence and intensity of the mood of a music clip	output-agreement	music mood classification
Phrase Detective [57]	players identify relationships between words and phrases in a short piece of text	output-agreement	natural language processing
Phylo [255]	players align colored blocks by moving them horizontally and inserting gaps	output-agreement	genome alignment

6.3   SUMMARY

This chapter addresses the question of design. Specifically, we reviewed different design considerations when creating a task as well as the use of game mechanisms to incentivize truthful outputs. The key points of this chapter include:

- The design of tasks involve five design decisions: the inclusion or exclusion of information, granularity, independence versus collaboration, incentives and quality control.

- Games with a purpose are simultaneously a way to motivate participation and a mechanism for ensuring truthful outputs.

PART III

Conclusion

CHAPTER 7

The Future of Human Computation

With the aim of being forward looking, this book both reviews existing work in human computation and related areas, as well as envisions future technologies for supporting human computation. Every chapter in this book offers some suggestions for future research; in this section, we will briefly mention three additional research directions that are not yet covered by the rest of the book.

7.1 RESEARCH DIRECTIONS

7.1.1 INTERWEAVING HUMAN AND MACHINE INTELLIGENCE

The premise behind human computation is that given the right conditions, an unprecedented number of human workers can be mobilized to solve problems that are difficult for machines to solve. However, as a resource, human computers are still typically more costly than machines. Hybrid solutions, involving both human and machine intelligence, may prove to be powerful for future human computation systems. There are many roles machine intelligence can play in human computation systems:

- **AI as optimizers.** There are many opportunities to leverage machine intelligence to help improve the accuracy and efficiency of human computation algorithms. We have seen in Chapter 3 some probabilistic models for aggregating outputs and inferring worker competence. As mentioned in Chapter 2, machine learning techniques, such as active learning, can help reduce the cost of human computation by choosing only informative queries to ask.

- **AI as enablers.** As human computation systems are built to handle increasingly complex tasks done by increasingly larger crowds (e.g., to generate disaster relief plan), we need to use machine intelligence to coordinate individuals, and to make sense of, organize and display information to workers. In other words, AI algorithms can be used to make *humans* compute better.

- **AI as workers.** For many tasks, machines actually outperform humans, both in terms of accuracy and speed. One can imagine future human computation systems to leverage both AI and humans as workers to perform different tasks they are better at.

An effective human computation system should be able to interweave machine and human capabilities seamlessly. This idea is not new; many research concepts familiar to the AI community, such as complementary computing [149], mixed-initiative systems [147] and interactive machine learning [113, 291, 359], address similar questions. There are already several human computation projects where we see hybrid solutions emerging. Systems like NELL (Never-Ending Language Learner) [53] are capable of very quickly sifting through huge amounts of data on the Web to find candidate facts, but have no means of verifying them; on the other hand, humans can easily verify the facts (via a game like Polarity) using past experience or external resources, e.g., search engine. CrowdPlan [193] asks workers to decompose high-level search queries (in the form of missions such as "I want to quit smoking") into a set of goals and transform the goals into search queries, leaving the actual search task to machines. Shahaf et al. [288] introduces the use of machine intelligence to manage human computers as a resource, by taking into account competencies, availabilities, and payment. Branson et al. [43] introduces a computer vision algorithm for classifying bird images, that upon facing uncertainties, asks Turkers to answer questions (e.g., "is the belly red") to refine the answer. To compute a crowd kernel (i.e., a similarity matrix over a set of objects), Tamuz et al. [311] uses an algorithm to adaptively select maximally informative triplets of objects to query for human similarity judgments, thereby approximately the true answer with fewer number of queries. We foresee these, and other AI technologies, to play an increasingly important role in human computation systems in the future.

7.1.2   FOSTERING LONG-TERM RELATIONSHIPS

Research so far assumes that workers come and go, and as a result, there are very few opportunities for the system to observe and learn about the characteristics of workers. However, having a stable pool of workers who return year after year is definitely possible, and in fact the case for many human computation systems. An important research question for the future is how to design human computation systems to promote and sustain long term relationships with workers. Several works have raised serious questions about the ethics of crowdsourcing work, including [111, 272, 292, 293]. For example, there exists power asymmetry in Mechanical Turk, where requesters can reject work without providing justification, thereby not only forbidding payment but hurting the future chances of work by damaging the workers' reputation. Other tasks are themselves malicious towards the workers (e.g., aimed to grab personal information, etc.) or badly designed (e.g., impossible to complete in the allotted amount of time) [292]. There is recent work on tools that address this power asymmetry by showing reviews of requesters, in terms of their communicativity, generosity, fairness and promptness [292]. In the future, the incorporation of manipulation-resistant reputation systems [1, 2, 33, 46, 84, 85, 119, 165, 247] might lead to greater transparency [4] and sense of trusts between the requesters and workers, which in turn, can encourage workers' long-term interaction with the system.

7.1.3   DESIGNING ORGANIZATIONS AND TASK MARKETS

In an article entitled "The Dawn of the E-lance Economy," Malone et al. [214] refers to business organizations as "mechanisms for coordination," which "guide the flow of work, materials, ideas and money." Mechanical Turk is one way of bringing together buyer and sellers of computational tasks, but that needs not be the only form of organization. In the future, human computation systems might dynamically assign roles (e.g., computer, checker, planner, task router, manager) to different workers, essentially creating dynamic, temporary organizations on the fly that are tailored to the particular computational problem at hand.

New forms of task markets may also emerge. Horton [144] describes two types of online labor market—"spot" market (e.g., oDesk, Elance) where buyer and sellers trade their goods at an agreed price and duration of time, and "contest" market (e.g., 99Designs, InnoCentive) where a buyer selects a winning solution out of a set of work performed by workers. One can imagine other types of task markets where requesters and workers are asked to express their needs and wants, and tasks can be allocated through a auction mechanism [59, 75, 277]. For any task markets, there will be opportunities to design smart brokers [14], intermediaries that help match requesters and workers by providing relevant information to both parties.

7.2   CONCLUSION

The purpose of this book is to provide a conceptual framework for thinking about human computation and an overview of the state-of-the-art work in the area. Being a relatively new research area, we foresee that this book will evolve as the field grows and concepts are clarified. Throughout the book, we outline the major aspects of human computation, including the core research questions related to algorithm-centric (Part I) and design-centric (Part II) aspects of human computation. Hosted on `http://humancomputation.com/book` is a list of references, datasets and tools, which may be of use to those conducting research in human computation. As the field progresses, we hope that this book will continue to serve as a useful resource for framing the problem of human computation, keeping track of what has been tackled and envisioning what is possible.

Bibliography

[1] L. A. Adamic, J. Zhang, E. Bakshy, and M. S. Ackerman. Knowledge sharing and yahoo answers: Everyone knows something. In *WWW*, pages 665–674, 2008. DOI: 10.1145/1367497.1367587 Cited on page(s) 74

[2] E. Agichtein, C. Castillo, D. Donato, A. Gionis, and G. Mishne. Finding high-quality content in social media. In *WSDM*, pages 183–194, 2008. DOI: 10.1145/1341531.1341557 Cited on page(s) 74

[3] N. Ailon, M. Charikar, and A. Newman. Aggregating inconsistent information: ranking and clustering. In *STOC*, 2005. DOI: 10.1145/1411509.1411513 Cited on page(s) 31

[4] G. Akerlof. The market for lemons: Quality uncertainty and the market mechanism. *Quarterly Journal of Economics*, 84:488—-500, 1970. DOI: 10.2307/1879431 Cited on page(s) 74

[5] A. Alchian and H. Demsetz. Production, information costs, and economic organization. *The American Economic Review*, 62(5):777–795, December 1972. DOI: 10.1109/EMR.1975.4306431 Cited on page(s) 62

[6] Amazon mechanical turk. https://www.mturk.com/mturk/welcome. Cited on page(s) 46

[7] Amazon mechanical turk requestor interface. https://requester.mturk.com. Cited on page(s) 47

[8] V. Ambati, S. Vogel, and J. Carbonell. Active learning and crowd-sourcing for machine translation. In *LREC*, 2010. Cited on page(s) 32, 35

[9] D. Ariely, U. Gneezy, G. Loewenstein, and N. Mazar. Large stakes and big mistakes. *Review of Economic Studies*, 75:1–19, 2009. DOI: 10.1111/j.1467-937X.2009.00534.x Cited on page(s) 61

[10] D. Ariely, E. Kamenica, and D. Prelec. Man's search for meaning: the case of legos. *Journal of Economic Behavior & Organization*, 67:671 – 677, 2008. DOI: 10.1016/j.jebo.2008.01.004 Cited on page(s) 61

[11] Dan Ariely, Wing Tung Au, Randall H. Bender, David V. Budescu, Christiane B. Dietz, Hongbin Gu, Thomas S. Wallsten, and Gal Zauberman. The effects of averaging subjective probability estimates between and within judges. *Journal of Experimental Psychology: Applied*, pages 130–147, 2000. DOI: 10.1037/1076-898X.6.2.130 Cited on page(s) 31

[12] S. Arora, E. Nyberg, and C.P. Rose. Estimating annotation cost for active learning in a multi-annotator environment. In *NAACL HLT Workshop on Active Learning for Natural Language Processing*, pages 18–26, 2009. DOI: 10.3115/1564131.1564136 Cited on page(s) 9

[13] R. Artstein and M. Poesio. Kappa^3 = alpha (or beta). Technical report, University of Essex, 2005. Cited on page(s) 28

[14] D. Autor. Wiring the labor market. *Journal of Economics Perspectives*, 15:25–40, 2001. DOI: 10.1257/jep.15.1.25 Cited on page(s) 75

[15] C. Avery, P. Resnick, and R. Zeckhauser. The market for evaluations. *American Economic Review*, 89(3):564–584, 1999. DOI: 10.1257/aer.89.3.564 Cited on page(s) 62

[16] M. Babaioff, M. Feldman, and N. Nisan. Combinatorial agency. In *ACM EC*, pages 18–28, 2006. DOI: 10.1145/1134707.1134710 Cited on page(s) 62

[17] F. Baker. *The Basics of Item Response Theory*. Heinemann, 1985. Cited on page(s) 29

[18] H. E. Bal, J. G. Steiner, and A. S. Tanenbaum. Programming languages for distributed computing systems. *ACM Computing Surveys*, 21(3), 1989. DOI: 10.1145/72551.72552 Cited on page(s) 17

[19] M. Balabanovic and Y. Shohan. Fab: content-based, collaborative recommendation. *CACM*, 40(3):66–72, 1997. DOI: 10.1145/245108.245124 Cited on page(s) 40

[20] J. Baldridge and M. Osborne. Active learning and the total cost of annotation. In *EMNLP*, 2004. Cited on page(s) 22

[21] S. Bangalore, G. Bordel, and G. Riccardi. Computing consensus translation from multiple machine translation systems. In *IEEE Workshop on Automatic Speech Recognition and Understanding*, 2001. DOI: 10.1109/ASRU.2001.1034659 Cited on page(s) 32

[22] A. Barabasi. The origin of bursts and heavy tails in human dynamics. *Nature*, 435:207–211, 2005. DOI: 10.1038/nature03459 Cited on page(s) 40

[23] G. Baron. The 106th christmas bird count. In *Audubon Society*, 2006. Cited on page(s) 51

[24] L. Barrington, D. O'Malley, D. Turnbull, and G. Lanckriet. User-centered design of a social game to tag music. In *KDD-HCOMP*, pages 7–10, 2009. DOI: 10.1145/1600150.1600152 Cited on page(s) 9, 69

[25] R. Barzilay. *Information Fusion for Multidocument Summarization*. PhD thesis, Columbia University, 2003. Cited on page(s) 32

[26] R. Barzilay and L. Lee. Learning to paraphrase: An unsupervised approach using multiple-sequence alignment. In *NAACL-HLT*, 2003. DOI: 10.3115/1073445.1073448 Cited on page(s) 30

[27] R. Barzilay, K. R. McKeown, and M. Elhadad. Information fusion in the context of multi-document summarization. In *ACL*, pages 550–557, 1999. DOI: 10.3115/1034678.1034760 Cited on page(s) 32

[28] R. Bell, Y. Koren, and C. Volinsky. Matrix factorization techniques for recommender systems. *IEEE Computer*, 42(8):30–37, 2009. DOI: 10.1109/MC.2009.263 Cited on page(s) 41

[29] P. N. Bennett, M. Chickering, and A. Mityagin. Learning consensus opinion: Mining data from a labeling game. In *WWW*, 2009. DOI: 10.1145/1526709.1526727 Cited on page(s) 64

[30] M. Bernstein, J.R. Brandt, R.C. Miller, and D.R. Karger. Crowds in two seconds: Enabling realtime crowd-powered interfaces. In *UIST*, 2011. Cited on page(s) 22

[31] M. Berstein, G. Little, R. Miller, B. Hartmann, M. Ackerman, D. Karger, D. Crowell, and K. Panovich. Soylent: A word processor with a crowd inside. In *UIST*, 2010. DOI: 10.1145/1866029.1866078 Cited on page(s) 17, 56, 60

[32] T. Bertin-Mahieux, D. Eck, F. Maillet, and P. Lamere. Autotagger: a model for predicting social tags from acoustic features on large music databases. *TASLP*, 37(2):115–135, 2008. DOI: 10.1080/09298210802479250 Cited on page(s) 8

[33] J. Bian, Y. Liu, E. Agichtein, and H. Zha. Finding the right facts in the crowd: Factoid question answering over social media. In *WWW*, pages 467–476, 2008. DOI: 10.1145/1367497.1367561 Cited on page(s) 74

[34] J. Bigham and A. Cavender. Evaluating existing audio captchas and an interface optimized for non-visual use. In *NIPS*, 2009. DOI: 10.1145/1518701.1518983 Cited on page(s) 49

[35] J.P. Bigham, C. Jayant, H. Ji, G. Little, A. Miller, R.C. Miller, R. Miller, A. Tatarowicz, B. White, S. White, and T. Yeh. Vizwiz: Nearly real-time answers to visual questions. In *UIST*, 2010. DOI: 10.1145/1866029.1866080 Cited on page(s) 21

[36] D. Billsus and M. J. Pazzani. Learning collaborative information filters. In *ICML*, pages 46–54, 1998. Cited on page(s) 41

[37] C. Boehm and G. Jacopini. Flow diagrams, turing machines and languages with only two formation rules. *Communications of the ACM*, 9(5):366–371, 1966. DOI: 10.1145/355592.365646 Cited on page(s) 17

[38] Boinc. http://boinc.berkeley.edu/. Cited on page(s) 4

[39] L. Bonetta. New citizens for the life sciences. *Cell*, 138:1043–1045, 2009. DOI: 10.1016/j.cell.2009.09.007 Cited on page(s) 52

[40] R. Bonney, C. Cooper, J. Dickinson, S. Kelling, T. Philips, K. Rosenberg, and J. Shirk. Citizen science: A developing tool for expanding science knowledge and scientific literacy. *BioScience*, 59:977–984, 2009. DOI: 10.1525/bio.2009.59.11.9 Cited on page(s) 54

[41] Boto. http://code.google.com/p/boto. Cited on page(s) 19

[42] B. Bouzy and T. Cazenave. Computer go: An ai oriented survey. *Artificial Intelligence*, 132(1):39–103, October 2001. DOI: 10.1016/S0004-3702(01)00127-8 Cited on page(s) 10

[43] S. Branson, C. Wah, B. Babenko, F. Schroff, P. Welinder, P. Perona, and S. Belongie. Visual recognition with humans in the loop. In *ECCV*, 2010. Cited on page(s) 52, 74

[44] C. Brodley and M. Friedl. Identifying mislabeled training data. *Journal of Artificial Intelligence Research*, 11:131–167, 1999. DOI: 10.1613/jair.606 Cited on page(s) 30

[45] D. Brossard, B. Lewenstein, and R. Bonney. Scientific knowledge and attitude change: the impact of a citizen science project. *International Journal of Science Education*, 27:1099–1121, 2005. DOI: 10.1080/09500690500069483 Cited on page(s) 54

[46] J. Brown and J. Morgan. Reputation in online auctions: The market for trust. *California Management Review*, 49(1):61–81, 2006. Cited on page(s) 74

[47] J. Burke, D. Estrin, M. Hansen, A. Parker, N. Ramanathan, S. Reddy, and M.B. Srivastava. Participatory sensing. In *ACM Sensys World Sensor Web Workshop*, 2006. Cited on page(s) 4

[48] R. Burke. Hybrid recommender systems: survey and experiments. *User Modeling and User-adapted Interaction*, 12(4):331–370, 2002. DOI: 10.1023/A:1021240730564 Cited on page(s) 41

[49] R. Burke, B. Mobasher, R. Bhaumik, and C. William. Segment-based injection attacks against collaborative filtering recommender systems. In *ICDM*, pages 577–580, 2005. DOI: 10.1109/ICDM.2005.127 Cited on page(s) 41

[50] C. Camerer, L. Babcock, G. Loewenstein, and R. Thaler. Labor supply of new york city cabdrivers: One day at a time. *The quarterly journal of economics*, 112(2):407–441, 1997. DOI: 10.1162/003355397555244 Cited on page(s) 61

[51] C. N. Cardamone, K. Schawinski, M. Sarzi, S. P. Bamford, N. Bennert, C. M. Urry, C. Lintott, W. C. Keel, J. Parejko, R. C. Nichol, D. Thomas, D. Andreescu, P. Murray, M. J. Raddick, A. Slosar, A. Szalay, and J. VandenBerg. Galaxy zoo green peas: Discovery of a class of compact extremely star-forming galaxies. *MNRAS*, 2009. DOI: 10.1111/j.1365-2966.2009.15383.x Cited on page(s) 51

[52] J. Carletta. Assessing agreement on classification tasks: the kappa statistics. *Computational Linguistics*, 22(2):249–254, 1996. Cited on page(s) 28

[53] A. Carlson, J. Betteridge, B. Kisiel, B. Settles, E.R. Hruschka Jr., and T.M. Mitchell. Towards an architecture for never-ending language learning. In *AAAI*, 2010. Cited on page(s) 74

[54] B. Carpenter. Multilevel bayesian models of categorical data annotation. Technical report, Alias-i, 2008. Cited on page(s) 29

[55] B. Carterette, P. N. Bennett, D. M. Chickering, and S. Dumais. Here or there: Preference judgments for relevance. In *ECIR*, pages 16–27, 2008. Cited on page(s) 18

[56] Catbook. http://bjoern.org/projects/catbook/. Cited on page(s) 60

[57] J. Chamberlain, M. Poesio, and U. Kruschwitz. Phrase detectives - a web-based collaborative annotation game. In *I-Semantics*, 2008. Cited on page(s) 69

[58] D. Chandler and A. Kepelner. Breaking monotony with meaning: Motivation in crowdsourcing markets. Technical report, MIT, 2010. working paper. Cited on page(s) 61

[59] K. Che. Design competition through multidimensional auctions. *RAND Journal of Economics*, 24(4):668–680, 1993. DOI: 10.2307/2555752 Cited on page(s) 75

[60] K. Chellapilla, K. Larson, P. Simard, and M. Czerwinski. Designing human friendly human interaction proofs. In *CHI*, 2005. DOI: 10.1145/1054972.1055070 Cited on page(s) 49

[61] Y. Chen, L. Fortnow, N. Lambert, D. Pennock, and J. Wortman. Complexity of combinatorial market makers. In *ACM EC*, 2008. DOI: 10.1145/1386790.1386822 Cited on page(s) 32

[62] Y. Chen and D. Pennock. Designing marekts for prediction. *AI Magazine*, 31(4):42–52, 2010. Cited on page(s) 32

[63] M. Chi. Two approaches to the study of experts' characteristics. In K. A. Ericsson, N. Charness, P. J. Feltovich, and R. R. Hoffman, editors, *The Cambridge handbook of expertise and expert performance*, pages 21–30. Cambridge University Press, 2006. Cited on page(s) 35

[64] L. Chilton, J. Horton, R. Miller, and S. Azenkot. Task search in a human computation market. In *KDD-HCOMP*, pages 1–9, 2010. DOI: 10.1145/1837885.1837889 Cited on page(s) 22, 40, 45

[65] R. Clemen and R. Winkler. Combining probability distributions from experts in risk analysis. *Risk Analysis*, 19(2):187–203, 1999. DOI: 10.1023/A:1006917509560 Cited on page(s) 31

[66] A. Cobham. Priority assignment in waiting line problems. *Journal of Operations Research Society of America*, 2:70–76, 1954. Cited on page(s) 40

[67] W. Cohen, R. Schapire, and Y. Singer. Learning to order things. *Journal of Artificial Intelligence*, 10:243–270, 1999. DOI: 10.1613/jair.587 Cited on page(s) 31

[68] Community collaborative rain, hail and snow network. `http://www.cocorahs.org/`. Cited on page(s) 52

[69] S. Cooper, F. Khatib, A. Treuille, J. Barbero, J. Lee, Michael Beenen, A. Leaver-Fay, D. Baker, Z. Popovic, and FoldIt Players. Predicting protein structures with a multiplayer online game. *Nature*, 466:756–760, August 2010. DOI: 10.1038/nature09304 Cited on page(s) 10, 59, 62, 69

[70] S. Cooper, A. Treuille, J. Barbero, A. Leaver-Fay, K. Tuite, F. Khatib, A. Snyder, M. Beenen, D. Salesin, D. Baker, Z. Popovic, and 000 Foldit players > 57. The challenge of designing scientific discovery games. In *FDG*, 2010. DOI: 10.1145/1822348.1822354 Cited on page(s) 52

[71] T. Cormen, C. Leiserson, and R. Rivest. *Introduction to Algorithms*. McGraw-Hill, 1999. Cited on page(s) 17

[72] Cornell lab of ornithology citizen science project. `http://cornellcitizenscience.org`. Cited on page(s) 54

[73] J.R. Corney, I. Kowalska, A.P. Jagadeesan, A. Lyn, Hugo Medellin, and W. Regli. Crowd-sourcing human problem solving strategy. In *CrowdConf*, 2010. Cited on page(s) 10, 62

[74] D. Cosley, D. Frankowski, L. Terveen, and J. Riedl. Suggestbot: using intelligent task routing to help people find work in wikiepdia. In *IUI*, 2007. DOI: 10.1145/1216295.1216309 Cited on page(s) 41

[75] P. Cramton, Y. Shoham, and R. Steinberg, editors. *Conbinatorial Auctions*. MIT Press, 2006. Cited on page(s) 75

[76] Cybertracker. `http://www.cybertracker.org/`. Cited on page(s) 52

[77] P. Dai, Mausam, and D. Weld. Decision-theoretic control of crowd-sourced workflows. In *AAAI*, 2010. Cited on page(s) 19

[78] P. Dai, Mausam, and D. Weld. Artificial intelligence for artificial artificial intelligence. In *AAAI*, 2011. Cited on page(s) 19

[79] A. Dawid and A. Skene. Maximum likelihood estimation of observer error-rates using the em algorithm. *Applied Statistics*, 28(1):20–28, 1979. DOI: 10.2307/2346806 Cited on page(s) 28

[80] L.T. DeCarlo and D.V. Cross. Sequential effects in magnitude scaling: models and theory. *Journal of Experimental Psychology: General*, 119:375–396, 1990. DOI: 10.1037/0096-3445.119.4.375 Cited on page(s) 59

[81] E. L. Deci, R. Koestner, and R. M. Ryan. A meta-analytic review of experiments examining the effects of extrinsic rewards on intrinsic motivation. *Psychological Bulletin*, 6(125):627–668, 1999. DOI: 10.1037/0033-2909.125.6.627 Cited on page(s) 61

[82] O. Dekel and O. Shamir. Good learners for evil teachers. In *ICML*, 2009. DOI: 10.1145/1553374.1553404 Cited on page(s) 30

[83] O. Dekel and O. Shamir. Vox populi: Collecting high-quality labels from a crowd. In *COLT*, 2009. Cited on page(s) 29

[84] C. Dellarocas. The digitization of word-of-mouth: promise and challenges of online reputation mechanisms. *Management Science*, 10(49):1407–1424, 2003. DOI: 10.1287/mnsc.49.10.1407.17308 Cited on page(s) 74

[85] C. Dellarocas, F. Dini, and G. Spagnolo. Designing reputation mechanisms. In G. Spagnolo N. Dimitri, G. Piga, editor, *Handbook of Procurement*. Cambridge University Press, 2007. Cited on page(s) 74

[86] C. Dellarocas and C. A. Wood. The sound of silence in online feedback: Estimating trading risks in the presence of reporting bias. *Management Science*, 54(3):460–476, 2008. DOI: 10.1287/mnsc.1070.0747 Cited on page(s) 29

[87] Deneme: a blog of experiments on amazon mechanical turk. http://groups.csail.mit.edu/uid/deneme. Cited on page(s) 21

[88] J. Deng, W. Dong, R. Socher, L.J. Li, and L. Fei-Fei. Imagenet: A large-scale hierarchical image database. In *CVPR*, 2009. Cited on page(s) 8

[89] P. Denning. The great principles of computing. *American Scientist*, September-October 2010. Cited on page(s) 1

[90] D. Dipalantino and M. Vojnovic. Corwdsourcing and all-pay auctions. In *ACM EC*, 2009. DOI: 10.1145/1566374.1566392 Cited on page(s) 62

[91] A. Doan, R. Ramakrishnan, and A. Y. Halevy. Mass collaboration systems on the world-wide web. *Communications of the ACM*, 54(4), 2011. DOI: 10.1145/1924421.1924442 Cited on page(s) 5

[92] P. Donmez and J. Carbonell. Proactive learning: Cost-sensitive active learning with multiple imperfect oracles. In *CIKM*, 2008. DOI: 10.1145/1458082.1458165 Cited on page(s) 38, 39

[93] P. Donmez and J. Carbonell. From active to proactive learning methods. In J. Koronacki, S. T. Wierzchon, Z. Ras, and J. Kacprzyk, editors, *Recent Advances in Machine Learning*. Springer: Studies in Computational Intelligence, 2009. Cited on page(s) 23

[94] P. Donmez, J. Carbonell, and J. Schneider. Efficiently leanring the accuracy of labeling sources for selective sampling. In *KDD*, 2009. DOI: 10.1145/1557019.1557053 Cited on page(s) 39

[95] P. Donmez, J. Carbonell, and J. Schneider. A probabilistic framework to learn from multiple annotators with time-varying accuracy. In *SDM*, 2010. Cited on page(s) 40

[96] C. Dwork, R. Kumar, M. Naor, and D. Sivakumar. Rank aggregation methods for the web. In *WWW*, 2001. DOI: 10.1145/371920.372165 Cited on page(s) 31

[97] E-bird. http://ebird.org/. Cited on page(s) 51

[98] N. Eagle. txteagle: Mobile crowdsourcing. *Lecture Notes in Computer Science: Internationalization, Design and Global Development*, 5623:447–456, 2009. DOI: 10.1007/978-3-642-02767-3_50 Cited on page(s) 39

[99] K. Ehrlich, C.Y. Lin, and V. Griffiths-Fisher. Searching for experts in the enterprise: combining text and social network analysis. In *GROUP*, 2007. DOI: 10.1145/1316624.1316642 Cited on page(s) 41

[100] Einstein@home. http://einstein.phys.uwm.edu/. Cited on page(s) 4

[101] A. Elo. *The Rating of Chessplayers, Past and Present*. Arco Publications, 1978. Cited on page(s) 31

[102] K. Eriksoon and B. Simpson. Emotional reactions to losing explain gender differences in entering a risky lottery. *Judgment and Decision*, 5(3):159–163, 2010. Cited on page(s) 46

[103] William S. Evans. *Information Theory and Noisy Computation*. PhD thesis, University of California at Berkeley, 1994. Cited on page(s) 21

[104] Evolution megalab. http://www.evolutionmegalab.org. Cited on page(s) 52

[105] Eyes of the reef. http://www.reefcheckhawaii.org/eyesofthereef.htm. Cited on page(s) 52

[106] R. Fagin, R. Kumar, M. Mahdian, D. Sivakumar, and E. Vee. Comparing partial rankings. *SIAM Journal of Discrete Mathematics*, 3(20):628–648, 2006. DOI: 10.1137/05063088X Cited on page(s) 31

[107] H. S. Farber. Reference-dependent preferences and labor supply: The case of new york city taxi drivers. *American Economic Review*, 98(3):1069–1082, 2008. Cited on page(s) 61

[108] S. Faridani, B. Hartmann, and P. Ipeirotis. What's the right price? pricing tasks for finishing on time. In *HCOMP*, 2011. Cited on page(s) 22

[109] E. Fehr and A. Falk. Psychological foundations of incentives. *European Economic Review*, 46(4-5):687 – 724, 2002. DOI: 10.1016/S0014-2921(01)00208-2 Cited on page(s) 61

[110] E. Fehr and L. Goette. Do workers work more if wages are high?: Evidence from a randomized field experiment. *American Economic Review*, 97(1):298–317, 2007. DOI: 10.1257/aer.97.1.298 Cited on page(s) 61

[111] A. Felstiner. Working the crowd: employment and labor law in the crowdsourcing industry. `http://ssrn.com/abstract=1593853`, 2011. Cited on page(s) 74

[112] P. Floreen, A. Kruger, and M. Spasojevic. Recruitment framework for participatory sensing data collection. In *Pervasive*, pages 138–155, 2010. DOI: 10.1007/978-3-642-12654-3_9 Cited on page(s) 36, 39

[113] J. Fogarty, D. Tan, A. Kapoor, and S. Winder. Cueflik: Interactive concept learning in image search. In *CHI*, 2008. DOI: 10.1145/1357054.1357061 Cited on page(s) 74

[114] Folding@home. `http://folding.stanford.edu/`. Cited on page(s) 4

[115] M. J. Franklin, D. Kossmann, T. Kraska, S. Ramesh, and R. Xin. Answering queries with crowdsourcing. In *SIGMOD*, 2011. DOI: 10.1145/1989323.1989331 Cited on page(s) 56

[116] R. Frederking and S. Nirenburg. Three heads are better than one. In *ANLP*, pages 95–100, 1994. Cited on page(s) 32

[117] B. S. Frey and R. Jegen. Motivation crowding theory. *Journal of Economic Surveys*, 5(15):589–611, 2001. DOI: 10.1111/1467-6419.00150 Cited on page(s) 61

[118] E. Friedman and P. Resnick. The social cost of cheap pseudonyms. *Journal of Economic Management Strategy*, 10(1):173–199, 2001. DOI: 10.1162/105864001300122476 Cited on page(s) 57

[119] E. Friedman, P. Resnick, and R. Sami. Manipulation-resistant reputation systems. In N. Nisan, T. Roughgarden, E. Tardos, and V. Vazirani, editors, *Algorithmic game theory*. Cambridge University Press, 2007. Cited on page(s) 74

[120] J. Froehlich, M.Y. Chen, I.E. Smith, and F. Potter. Voting with your feet: An investigative study of the relationship between place visit behavior and preference. In *Ubicomp*, 2006. DOI: 10.1007/11853565_20 Cited on page(s) 39

[121] D. Gale and L. S. Shapley. College admissions and the stability of marriage. *The American Mathematical Monthly*, 1(69):9–15, 1962. DOI: 10.2307/2312726 Cited on page(s) 38

[122] F. Garcin, B. Faltings, R. Jurca, and N. Joswig. Rating aggregation in collaborative filtering systems. In *the third ACM Conference on Recommender Systems*, pages 349–352, 2009. DOI: 10.1145/1639714.1639785 Cited on page(s) 31

[123] M. R. Garey and D. S. Johnson. *Computers and Intractibility: A guide to the theory of NP-completeness*. W.H. Freeman, 1979. Cited on page(s) 10

[124] A. Ghosh and P. Hummel. A game-theoretic analysis of rank-order mechanisms for user-generated content. In *ACM EC*, 2011. DOI: 10.1145/1993574.1993603 Cited on page(s) 61

[125] A. Ghosh, S. Kale, and P. McAfee. Who moderates the moderators? crowdsourcing abuse detection in user-generated content. In *ACM EC*, 2011. DOI: 10.1145/1993574.1993599 Cited on page(s) 31

[126] A. Ghosh and P. McAfee. Incentivizing high-quality user-generated content. In *WWW*, 2011. DOI: 10.1145/1963405.1963428 Cited on page(s) 61

[127] A. Gionis, H. Mannila, and P. Tsaparas. Clustering aggregation. In *ICDE*, pages 341–352, 2005. DOI: 10.1145/1217299.1217303 Cited on page(s) 31

[128] U. Gneezy and A. Rustichini. Pay enough or don't pay at all. *Quarterly journal of economics*, 115:791–810, 2000. DOI: 10.1162/003355300554917 Cited on page(s) 61

[129] A. Goder and V. Filkov. Consensus clustering algorithms: Comparison and refinement. In *SIAM Ninth Workshop on Algorithm Engineering and Experiments (ALENEX)*, 2008. Cited on page(s) 31

[130] Google image labeler. http://images.google.com/imagelabeler/help.html. Cited on page(s) 8, 50

[131] D. A. Grier. *When Computers were Human*. Princeton University Press, 2005. Cited on page(s) 1, 2, 51

[132] G. Gutin and A. Punnen. *The Traveling Salesman Problem and Its Variations*. Springer, 2006. Cited on page(s) 10

[133] S. Hacker and L. von Ahn. Eliciting user preferences with an online game. In *CHI*, pages 1207–1216, 2009. DOI: 10.1145/1518701.1518882 Cited on page(s) 18, 31, 64

[134] J. W. Hatfield and S. D. Kominers. Matching in networks with bilateral contracts. In *ACM EC*, pages 119–120, 2010. DOI: 10.1145/1807342.1807361 Cited on page(s) 38

[135] R. Herbrich, T. Minka, and T. Graepel. Trueskilltm: A bayesian skill rating system. In *NIPS*, pages 569–576, 2007. Cited on page(s) 31

[136] A. Herdagdelen and M. Baroni. The concept game: Better commonsense knowledge extraction by combining text mining and a game with a purpose. In *AAAI Fall Symposium on Commonsense Knowledge*, 2010. Cited on page(s) 68

[137] J. L. Herlocker, J. Konstan, L. Terveen, and J. Riedl. Evaluating collaborative filtering recommender systems. *ACM Transactions on Information Systems*, 22(1):5–53, 2004. DOI: 10.1145/963770.963772 Cited on page(s) 41

[138] B. E. Hermalin. Toward an economic theory of leadership: leading by example. *The American Economic Review*, 88(5):1188–1206, December 1998. DOI: 10.2139/ssrn.15570 Cited on page(s) 62

[139] P. Heymann and H. Garcia-Molina. Human processing. Technical report, Standard University, July 2010. Cited on page(s) 21

[140] A. V. Hill. An experimental comparison of human schedulers and heuristic algorithms for the traveling salesman problem. *Journal of Operations Management*, 2(4):215–223, August 1982. DOI: 10.1016/0272-6963(82)90010-9 Cited on page(s) 10

[141] C.J. Ho, T.H. Chang, J.C. Lee, J. Hsu, and K.T. Chen. Kisskissban: a competitive human computation game for image annotation. In *Proceedings of the first Workshop on Human Computation*, pages 11–14, 2009. DOI: 10.1145/1600150.1600153 Cited on page(s) 68

[142] J. Hodge and R. E. Klima. *The Mathematics of Voting and Elections: A Hands-On Approach*. American Mathematics Society, 2000. Cited on page(s) 31

[143] M. Hoffman, D. Blei, and P. Cook. Easy as cba: A simple probabilistic model for tagging music. In *ISMIR*, pages 369–374, 2009. Cited on page(s) 8

[144] J. Horton. Online labor markets. In *the 6th Workshop on Internet and Network Economics (WINE)*, 2010. DOI: 10.1007/978-3-642-17572-5_45 Cited on page(s) 75

[145] J. Horton and L. Chilton. The labor economics of paid crowdsourcing. In *EC*, 2010. DOI: 10.1145/1807342.1807376 Cited on page(s) 61

[146] J. Horton, D. Rand, and R. Zeckhauser. The online laboratory. In *WINE*, 2010. Cited on page(s) 46, 62

[147] E. Horvitz. Reflections on challenges and promises of mixed-initiative interaction. *AAAI Magazine*, 28, 2007. Cited on page(s) 74

[148] E. Horvitz, J. Apacible, and P. Koch. Busybody: Creating and fielding personalized models of the cost of interruption. In *CSCW*, 2004. DOI: 10.1145/1031607.1031690 Cited on page(s) 39

[149] E. Horvitz and T. Paek. Complementary computing: Policies for transferring callers from dialog systems to human receptionists. *User Modeling and User Adapted Interaction*, 17, 2007. DOI: 10.1007/s11257-006-9026-1 Cited on page(s) 6, 74

[150] E. Huang, H. Zhang, D. Parkes, K. Gajos, and Y. Chen. Toward automatic task design: A progress report. In *Proceedings of the second Human Computation Workshop*, 2010. DOI: 10.1145/1837885.1837908 Cited on page(s) 23, 55, 60

[151] Y. Huang. *Mixed-initiative clustering*. PhD thesis, Carnegie Mellon University, 2010. Cited on page(s) 31

[152] Human computation supplementary resources. http://humancomputation.com/book. Cited on page(s) 46

[153] L. Hurwicz. On informationally decentralized systems. In C.B. McGuire and R. Radner, editors, *Decision and Organization: a Volume in Honor of Jacob Marshak*, pages 297–336. North-Holland, 1972. Cited on page(s) 63

[154] Igda causal games white paper 2008–2009. http://archives.igda.org/casual/IGDA_Casual_Games_White_Paper_2008.pdf. Cited on page(s) 49, 50

[155] Imagenet large scale visual recognition challenge 2010. http://www.image-net.org/challenges/LSVRC/2010/index. Cited on page(s) 8

[156] P. Ipeirotis. Analyzing the amazon mechanical turk marketplace. *XRDS*, 17(2), 2010. DOI: 10.1145/1869086.1869094 Cited on page(s) 45, 46, 61

[157] P. Ipeirotis. Demographics of mechanical turk. Technical report, New York University, 2010. Working Paper. Cited on page(s) 47

[158] P. Ipeirotis, F. Provost, and J. Wang. Quality management on amazon mechanical turk. In *HCOMP*, 2010. DOI: 10.1145/1837885.1837906 Cited on page(s) 26, 28, 29

[159] M. O. Jackson. Mechanism theory. In U. Derigs, editor, *Encyclopedia of Life Support Systems*. EOLSS Publishers, 2003. Cited on page(s) 62

[160] S. Jain, Y. Chen, and D. Parkes. Designing incentives for online question and answers forums. In *EC*, pages 129–138, 2009. DOI: 10.1145/1566374.1566393 Cited on page(s) 61

[161] S. Jain and D. Parkes. A game-theoretic analysis of games with a purpose. In *WINE '08: Proceedings of the 4th International Workshop on Internet and Network Economics*, pages 342–350, 2008. DOI: 10.1007/978-3-540-92185-1_40 Cited on page(s) 67

[162] S. Jain and D. Parkes. Combinatorial agency of threshold functions. In *ACM EC Workshop on social computing and user generated content*, 2011. Cited on page(s) 62

[163] S. Jayaraman and A. Lavie. Multi-engine machine translation guided by explicit word matching. In *EAMT*, 2005. DOI: 10.3115/1225753.1225779 Cited on page(s) 32

[164] T. Joachims. Optimizing search engines using clickthrough data. In *KDD*, 2002. DOI: 10.1145/775047.775067 Cited on page(s) 31

[165] R. Jurca and B. Faltings. An incentive-compatible reputation mechanism. In *IEEE Conference on E-Commerce*, pages 285–292, 2003. DOI: 10.1145/860575.860778 Cited on page(s) 74

[166] R. Jurca and B. Faltings. Mechanisms for making crowds truthful. *Journal of Artificial Intelligence Research*, 34(34):209–253, 2009. Cited on page(s) 32

[167] L. Kaelbling. *Learning in Embedded Systems*. The MIT Press, Cambridge, MA, 1993. Cited on page(s) 39

[168] L. Kaelbling, M. Littman, and A. Moore. Reinforcement learning: A survey. *Journal of Artificial Intelligence Research*, 4:237–285, 1996. Cited on page(s) 39

[169] A. Kapoor and E. Horvitz. On discarding, caching, and recalling samples in active learning. In *UAI*, 2007. Cited on page(s) 39

[170] A. Kapoor and E. Horvitz. Experience sampling for building predictive user models: A comparative study. In *CHI*, 2008. DOI: 10.1145/1357054.1357159 Cited on page(s) 39

[171] G. Karabatsos and W. Batchelder. Markov chain estimation for test theory without an answer key. *Psychometrika*, 68(3):373–389, 2003. DOI: 10.1007/BF02294733 Cited on page(s) 29

[172] M. Kearn, S. Suri, and N. Montfort. An experimental study of the coloring problem on human subject networks. *Science*, 313(5788):824–827, August 2006. DOI: 10.1126/science.1127207 Cited on page(s) 4, 10, 18

[173] A. J. Kim. *Community building on the web: secret strategies for successful online communities*. Peachpit Press, 2000. Cited on page(s) 55

[174] Y.E. Kim, E. Schmidt, and L. Emelle. Moodswings: A collaborative game for music mood label collection. In *ISMIR*, pages 231–236, 2008. Cited on page(s) 69

[175] A. Kittur. Crowdsourcing, collaboration and creativity. *XRDS*, 17(2):22 – 26, 2010. DOI: 10.1145/1869086.1869096 Cited on page(s) 59, 60, 62

[176] A. Kittur, E. Chi, and B. Suh. Crowdsourcing user studies with mechanical turk. In *CHI*, 2008. DOI: 10.1145/1357054.1357127 Cited on page(s) 60, 62

[177] A. Kittur and R. Kraut. Harnessing the wisdom of crowds in wikipedia: Quality through coordination. In *CHI*, 2008. DOI: 10.1145/1460563.1460572 Cited on page(s) 5

[178] A. Kittur, B. Smus, and R. Kraut. Crowdforge: Crowdsourcing complex work. Technical report, Carnegie Mellon University, 2011. DOI: 10.1145/1979742.1979902 Cited on page(s) 19, 62

[179] A. Kittur, B. Suh, B. Pendleton, and E. Chi. He says, she says: Conflict and coordination in wikipedia. In *CHI*, 2007. Cited on page(s) 5

[180] G. W. Klau, N. Lesh, J. Marks, and M. Mitzenmacher. Human-guided tabu search. In *AAAI*, pages 41–47, 2002. DOI: 10.1007/s10732-009-9107-5 Cited on page(s) 10, 11

[181] G. W. Klau, N. Lesh, J. Marks, M. Mitzenmacher, and G. T. Schafer. The hugs platform: a toolkit for interactive optimization. In *Advanced Visual Interfaces*, pages 324–330, 2002. DOI: 10.1145/1556262.1556314 Cited on page(s) 10, 11

[182] D. Knuth. *The Art of Computer Programming: Second Edition*. Addison-Wesley, 1973. Cited on page(s) 16

[183] R. Kowalski. Algorithm = logic + control. *Communications of the ACM*, 22(7):424–436, July 1979. DOI: 10.1145/359131.359136 Cited on page(s) 17

[184] R. E. Kraut and P. Resnick. *Evidence-based social design: Mining the social sciences to build online communities*. MIT Press, 2011. Cited on page(s) 55, 61

[185] A. Kulkarni and M. Can. Turkomatic: Automatic recursive task design for mechanical turk. In *CHI Work-in-Progress*, 2011. Cited on page(s) 19, 59

[186] Y.L. Kuo, K.Y. Chiang, C.W. Chan, J.C. Lee, R. Wang, E. Shen, and J. Hsu. Community-based game design: Experiments on social games for commonsense data collection. In *KDD Human Computation Workshop (HCOMP 2009)*, 2009. DOI: 10.1145/1600150.1600154 Cited on page(s) 45, 62

[187] S.K. Lam and J. Riedl. Shilling recommender systems for fun and profit. In *WWW*, pages 393–402, 2004. DOI: 10.1145/988672.988726 Cited on page(s) 41

[188] E. Law, P. Bennett, and E. Horvitz. The effects of choice in routing relevance judgments. In *SIGIR*, 2011. Cited on page(s) 40

[189] E. Law, B. Settles, and T. Mitchell. Learning to tag from open vocabulary labels. In *ECML*, 2010. DOI: 10.1007/978-3-642-15883-4_14 Cited on page(s) 30, 51, 67

[190] E. Law and L. von Ahn. Input-agreement: A new mechanism for data collection using human computation games. In *CHI*, pages 1197–1206, 2009. DOI: 10.1145/1518701.1518881 Cited on page(s) 9, 25, 51, 65, 69

[191] E. Law, L. von Ahn, R. Dannenberg, and M. Crawford. Tagatune: A game for music and sound annotation. In *ISMIR*, 2007. Cited on page(s) 64

[192] E. Law, K. West, M. Mandel, M. Bay, and S. Downie. Evaluation of algorithms using games: The case of music tagging. In *ISMIR*, 2009. Cited on page(s) 9

[193] E. Law and H. Zhang. Towards large-scale collaborative planning: Answering high-level search queries using human computation. In *AAAI*, 2011. Cited on page(s) 59, 74

[194] J. Lee and B. Hoh. Sell your experiences: A market mechanism based incentive for participatory sensing. In *IEEE PerCom*, 2010. DOI: 10.1109/PERCOM.2010.5466993 Cited on page(s) 62

[195] D. Levy and M. Newborn. *How Computers Play Chess*. W.H. Freeman, 1990. Cited on page(s) 10

[196] D. Lewis and J. Catlett. Heterogeneous uncertainty sampling for supervised learning. In *ICML*, pages 148–156, 1994. Cited on page(s) 23

[197] Lhca@home. `http://lhcathome.cern.ch/`. Cited on page(s) 4

[198] C.Y. Lin, K. Ehrlich, V. Griffiths-Fisher, and C. Desforges. Smallblue: people mining for expertise search. *IEEE Multimedia*, 2008. DOI: 10.1109/MMUL.2008.17 Cited on page(s) 41

[199] G. Linden, B. Smith, and J. York. Amazon.com recommendations: item-to-item collaborative filtering. *IEEE Internet Computing*, 7(1):76–80, 2003. DOI: 10.1109/MIC.2003.1167344 Cited on page(s) 41

[200] C. Lintott, K. Schawinski, S. Bamford, A. Slosar, K. Land, D. Thomas, E. Edmondson, K. Masters, R. Nichol, J. Raddick, A. Szalay, D. Andreescu, P. Murray, and J. Vandenberg. Galaxy zoo 1 : Data release of morphological classifications for nearly 900,000 galaxies. *MNRAS*, 2010. DOI: 10.1111/j.1365-2966.2010.17432.x Cited on page(s) 51

[201] C.J. Lintott, K. Schawinski, A. Slosar, K. Land, S. Bamford, D. Thomas, M.J. Raddick, R. Nichol, A.S. Szalay, D. Andreescu, P. Murray, and J. Vandenberg. Galaxy zoo: Morphologies derived from visual inspection of galaxies from the sloan digital sky survey. *MNRAS*, 389:1179, 2008. DOI: 10.1111/j.1365-2966.2008.13689.x Cited on page(s) 51

[202] G. Little, L. Chilton, M. Goldman, and R. Miller. Exploring iterative and parallel human computation processes. In *HCOMP*, 2010. DOI: 10.1145/1837885.1837907 Cited on page(s) 32, 59

[203] G. Little, L. Chilton, M. Goldman, and R. Miller. Turkit: Human computation algorithms on mechanical turk. In *UIST*, 2010. DOI: 10.1145/1866029.1866040 Cited on page(s) 59

[204] G. Little, L. Chilton, R. Miller, and M. Goldman. Turkit: Tools for iterative tasks on mechanical turk. In *HCOMP*, 2009. DOI: 10.1145/1600150.1600159 Cited on page(s) 17, 19, 20

[205] D. Lizotte, O. Madani, and R. Greiner. Budgeted learning of naive-bayes classifiers. In *UAI*, 2003. Cited on page(s) 23

[206] R.D. Luce and H. Raiffa. *Games and Decisions: Introduction and Critical Survey*. Wiley, 1957. Cited on page(s) 38

[207] G. Lugosi. Learning with an unreliable teacher. *Pattern Recognition*, 25(1):79–87, 1992. DOI: 10.1016/0031-3203(92)90008-7 Cited on page(s) 30

[208] C. Lundmark. Bioblitz: Getting into backyard biodiversity. *Bioscience*, 53:329, 2003. DOI: 10.1641/0006-3568(2003)053[0329:BGIBB]2.0.CO;2 Cited on page(s) 52

[209] I. Lynce and J. Ouaknine. Sudoku as a sat problem. In *9th Symposium on Artificial Intelligence and Mathematics*, 2006. Cited on page(s) 10

[210] C. Macdonald and I. Ounis. Voting for candidates: adapting data fusion techniques for an expert search task. In *CIKM*, pages 387–396, 2006. DOI: 10.1145/1183614.1183671 Cited on page(s) 41

[211] J.N. MacGregor and T. Ormerod. Human performance on the traveling salesman problem. *Perception & Psychophysics*, 58(4):527–539, 1996. DOI: 10.3758/BF03213088 Cited on page(s) 10

[212] Magnatagatune. http://tagatune.org/Magnatagatune.html. Cited on page(s) 9

[213] T. Malone. Heuristics for designing enjoyable user interfaces. In *CHI*, pages 63–68, 1982. DOI: 10.1145/800049.801756 Cited on page(s) 63

[214] T.W. Malone and R.J. Laubacher. The dawn of the e-lance economy. *Harvard Business Review*, 76(5):144 – 152, 1998. Cited on page(s) 5, 74

[215] M. Mandel and D. Ellis. A web-based game for collecting music metadata. *Journal of New Music Research*, 37(2):151–165, 2009. DOI: 10.1080/09298210802479300 Cited on page(s) 9, 64

[216] A. Mao, D. Parkes, A. Procaccia, and H. Zhang. Human computation and multiagent systems: An algorithmic perspective. In *working paper*, 2011. Cited on page(s) 18

[217] A. Marcus, E. Wu, D. Karger, S. Madden, and R. Miller. Crowdsourced databases: Query processing with people. In *CIDR*, 2011. Cited on page(s) 56

[218] M. Marge, S. Banerjee, and A. I. Rudnicky. Using the amazon mechanical turk for transcription of spoken language. In *ICASSP*, 2010. DOI: 10.1109/ICASSP.2010.5494979 Cited on page(s) 8

[219] W. Mason and S. Suri. Conducting behavioral research on amazon's mechanical turk. In *Social Science Research Network Working Paper Series*, 2010. Cited on page(s) 46, 47

[220] W. Mason and D. Watts. Financial incentives and the "performance of the crowds". In *Proceedings of the first Human Computation Workshop*, 2009. DOI: 10.1145/1600150.1600175 Cited on page(s) 61

[221] W. Mason and D. Watts. Collaborative problem solving in networks. In Submission, 2011. Cited on page(s) 60

[222] E. Matusov, N. Ueffing, and H. Ney. Computing consensus translation from multiple machine translation systems using enhanced hypotheses alignment. In *EACL*, pages 33–40, 2006. Cited on page(s) 32

[223] S.T. Maytal, Prem Melville, and F. Provost. Active feature-value acquisition for model induction. *Management Science*, 55(4):664–684, 2009. DOI: 10.1287/mnsc.1080.0952 Cited on page(s) 39

[224] S.T. Maytal and F. Provost. Decision-centric active learning of binary-outcome models. *Information Systems Research*, 18(1):1–19, 2007. DOI: 10.1287/isre.1070.0111 Cited on page(s) 39

[225] D.W. McDonald and M.S. Ackerman. Expertise recommender: A flexible recommendation architecture. In *CSCW*, pages 231–240, 2000. DOI: 10.1145/358916.358994 Cited on page(s) 41

[226] C. McMillen and M. Veloso. Thresholded rewards: acting optimally in timed, zero-sum games. In *AAAI*, 2007. Cited on page(s) 21

[227] C. McMillen and M. Veloso. Unknown rewards in finite-horizon domains. In *AAAI*, 2008. Cited on page(s) 21

[228] P. Milgrom. Package auctions and package exchanges. *Econometrica*, 75(4):935–966, 2007. DOI: 10.1111/j.1468-0262.2007.00778.x Cited on page(s) 38

[229] N. Miller, P. Resnick, and R. Zeckhauser. Eliciting honest feedback: the peer prediction method. *Management Science*, 9(51):1359–1373, 2005. DOI: 10.1287/mnsc.1050.0379 Cited on page(s) 32, 61

[230] Mirex. http://www.music-ir.org/mirex/wiki/MIREX_HOME. Cited on page(s) 9

[231] B. Moldovanu and A. Sela. The optimal allocation of prizes in contests. *American economic review*, 91:542–558, June 2001. DOI: 10.1257/aer.91.3.542 Cited on page(s) 62

[232] B. Moldovanu and A. Sela. Contest architecture. *Journal of economic theory*, 126(1):70–96, January 2006. DOI: 10.1016/j.jet.2004.10.004 Cited on page(s) 62

[233] G. Mori and J. Malik. Recognizing objects in adversarial clutter: Breaking a visual catpcha. In *CPVR*, 2003. DOI: 10.1109/CVPR.2003.1211347 Cited on page(s) 49

[234] M. Motoyama, K. Levchenko, C. Kanich, D. McCoy, G. M. Voelker, and S. Savage. Re: Captchas – understanding captcha-solving services in an economic context. In *USENIX Security Symposium*, 2010. Cited on page(s) 49

[235] M. Mozer, H. Pashler, M. Wilder, R. Lindsey, M. Jones, and M. Jones. Decontaminating human judgments by removing sequential dependencies. In *NIPS*, 2010. Cited on page(s) 59

[236] R. Munro. Crowdsourced translation for emergency response in haiti: the global collaboration of local knowledge. In *AMTA Workshop on Collaborative Crowdsourcing for Translation*, 2010. Cited on page(s) 54

[237] K. K. Nam, M. S. Ackerman, and L. A. Adamic. Questions in, knowledge in? a study of naver's question answering community. In *CHI*, 2009. DOI: 10.1145/1518701.1518821 Cited on page(s) 61

[238] G. Nemhauser, L. Wolsey, and M. Fisher. An analysis of approximations for maximizing submodular set functions. *Mathematical Programming*, 14(1):265–294, 1978. DOI: 10.1007/BF01588971 Cited on page(s) 37

[239] J. Nielsen. *Usability Engineering*. Morgan Kaufmann, 1993. Cited on page(s) 45

[240] N. Nisan. Introduction to mechanism design (for computer scientists). In N. Nisan, T. Roughgarden, E. Tardos, and V. Vazirani, editors, *Algorithmic Game Theory*. Cambridge University Press, 2007. Cited on page(s) 62, 63

[241] D.W. North. A tutorial introduction to decision theory. *IEEE Transactions on Systems Science and Cybernetics*, 4(3):200–210, 1968. DOI: 10.1109/TSSC.1968.300114 Cited on page(s) 38, 39

[242] C. Notredame. Recent progresses in multiple sequence alignment: a survey. *Pharmacogenomics*, 31(1):131–144, 2002. DOI: 10.1517/14622416.3.1.131 Cited on page(s) 10

[243] C. Notredame. Pulsar discovery by global volunteer computing. *Science*, 329(5997):1305, September 2010. DOI: 10.1126/science.1195253 Cited on page(s) 4

[244] P. O'Keefe. *Managing online forums: everything you need to know to create and run successful community discussion boards*. AMACOM, 2008. Cited on page(s) 55

[245] F. Olsson. A literature survey of active machine learning in the context of natural language processing. Technical report, Swedish Institute of Computer Science, 2009. Cited on page(s) 22

[246] G. Paolacci, J. Chandler, and P. Ipeirotis. Running experiments on amazon mechanical turk. *Judgment and Decision*, 5(5):411–419, 2010. Cited on page(s) 46

[247] T. Papaioannou and G. D. Stamoulis. An incentives' mechanism promoting truthful feedback in peer-to-peer systems. In *ACM CCGRID Workshop on Global P2P Computing*, 2005. DOI: 10.1109/CCGRID.2005.1558565 Cited on page(s) 74

[248] A. Parameswaran and N. Polyzotis. Answering queries using humans, algorithms and databases. In *CIDR*, 2011. Cited on page(s) 56

[249] A. Parameswaran, A. Das Sarma, H. Garcia-Molina, N. Polyzotis, and Jennifer Widom. Human-assisted graph search: It's okay to ask questions. Technical report, Stanford University, 2011. Cited on page(s) 10

[250] A. Parducci. The relativism of absolute judgements. *Scientific American*, 219(6):84–90, 1968. DOI: 10.1038/scientificamerican1268-84 Cited on page(s) 59

[251] A. Parducci. Category ratings and the relational character of judgment. In H.G. Geissler, editor, *Modern issues in perception, volume 1*, pages 262–282. Elsevier Science Publishers, 1991. Cited on page(s) 59

[252] R. Patz, B. Junker, and M. Johnson. The hierarchical rater model for rated test items and its application to large-scale educational assessment data. *Journal of Educational and Behavioral Statistics*, 27(4):341–381, 2002. DOI: 10.3102/10769986027004341 Cited on page(s) 29

[253] E. Paulos. Designing for doubt: Citizen science and the challenge of change. In *Engaging Data: First International Forum on the Application and Management of Personal Electronic Information*, 2009. Cited on page(s) 52

[254] D. Pennock and R. Sami. Computational aspects of prediction marekts. In N. Nisan, T. Roughgarden, E. Tardos, and V. Vazirani, editors, *Algorithmic game theory*. Cambridge University Press, 2007. Cited on page(s) 32

[255] Phylo – a human computing framework for comparative genomics. http://phylo.cs.mcgill.ca/. Cited on page(s) 10, 69

[256] K. Pipineni, S. Roukos, T. Ward, and W. Zhu. Bleu: a method for automatic evaluation of machine translation. In *ACL*, pages 311–318, 2002. DOI: 10.3115/1073083.1073135 Cited on page(s) 32

[257] N. Pippenger. Reliable computation by formulas in the presence of noise. *IEEE Transactions on Information Theory*, 34(2):194–197, 1988. DOI: 10.1109/18.2628 Cited on page(s) 21

[258] D. M. Powazek. *Design for community: the art of connecting real people in virtual places*. Pearson Technology, 2002. Cited on page(s) 55

[259] J. Preece. *Online communities: designing usability and supporting sociability*. John Wiley & Sons, 2000. Cited on page(s) 55

[260] Primegrid. `http://www.primegrid.com/`. Cited on page(s) 4

[261] M.J. Raddick, G. Bracey, P.L. Gay, C.J. Lintott, P. Murray, K. Schawinski, A.S. Szalay, and J. Vandenberg. Galaxy zoo: Exploring the motivations of citizen science volunteers. *MNRAS*, 2010. DOI: 10.3847/AER2009036 Cited on page(s) 51, 54

[262] F. Radlinski and T. Joachims. Query chains: learning to rank from implicit feedback. In *KDD*, 2005. DOI: 10.1145/1081870.1081899 Cited on page(s) 31

[263] H. Raghavan, O. Madani, and R. Jones. Active learning with feedback on both features and instances. *Journal of Machine Learning Research*, 7:1655–1686, 2006. Cited on page(s) 23

[264] V. Raykar, S. Yu, L. Zhao, G. Valadez, C. Florin, L. Bogoni, and L. Moy. Learning from crowds. *Journal of Machine Learning Research*, 11:1297–1322, 2010. Cited on page(s) 29, 30

[265] V.C. Raykar, S. Yu, L.H. Zhao, A. Jerebko, C. Florin, G.H. Valadez, L. Bogoni, and L. Moy. Supervised learning from multiple experts: Whom to trust when everyone lies a bit. In *ICML*, 2009. DOI: 10.1145/1553374.1553488 Cited on page(s) 29, 30

[266] S. Reiss. Multifaceted nature of intrinsic motivation: the theory of 16 basic desires. *Review of General Psychology*, 8:179–193, 2004. DOI: 10.1037/1089-2680.8.3.179 Cited on page(s) 61

[267] P. Resnick, I. Neophytos, P. Bergstrom, S. Mitesh, and J. Reidl. Grouplens: An open architecture for collaborative filtering of netnews. In *CSCW*, 1994. DOI: 10.1145/192844.192905 Cited on page(s) 41

[268] K. Ristovski, D. Das, V. Ouzienko, Y. Guo, and Z. Obradovic. Regression learning with multiple noisy oracles. In *ECAI*, 2010. Cited on page(s) 30

[269] M. Rodriguez and J. Bollen. An algorithm to determine peer reviewers. In *CIKM*, 2008. DOI: 10.1145/1458082.1458127 Cited on page(s) 40

[270] K. Romney. Culture consensus as a statistical model. *Current Anthropology*, 40:103–115, 1999. DOI: 10.1086/200063 Cited on page(s) 27

[271] K. Romney, S. Weller, and W. Batchelder. Culture as consensus: A theory of culture and informant accuracy. *American Anthropology*, 88(2):313–338, 1986 DOI: 10.1525/aa.1986.88.2.02a00020 Cited on page(s) 27, 29

[272] J. Ross, L. Irani, M.S. Silberman, A. Zaldivar, and B. Tomlinson. Who are the crowdworkers? shifting demographics in mechanical turk. In *CHI Work-in-Progress*, pages 2863–2872, 2010. DOI: 10.1145/1753846.1753873 Cited on page(s) 46, 47, 74

[273] A.-V.I. Rosti, N.F. Ayan, B. Xiang, S. Matsoukas, R. Schwartz, and B.J. Dorr. Combining outputs from multiple machine translation systems. In *NAACL-HLT*, pages 228–235, 2007. Cited on page(s) 32

[274] A. E. Roth. The economics of matching: stability and incentives. *Mathetmatics of Operations Research*, 4(7):617–628, 1982. DOI: 10.1287/moor.7.4.617 Cited on page(s) 38

[275] A. E. Roth. The economist as engineer: game theory, experimentation, and computation as tools for design economics. *Econometrica*, 4(1):1341–1378, 2002. DOI: 10.1111/1468-0262.00335 Cited on page(s) 38

[276] N. Roy and A. McCallum. Toward optimal active learning through sampling estimation of error reduction. In *ICML*, pages 441–448, 2001. Cited on page(s) 23

[277] T. Sandholm. Expressive commercse and its application to sourcing: how we conducted $35 billion of generalized combinatorial auctions. *AI Magazine*, 28(3):45–58, 2007. DOI: 10.1145/1282100.1282165 Cited on page(s) 75

[278] B. Sarwar, G. Karypis, J. Konstan, and J. Reidl. Item-based collaborative filtering algorithms. In *WWW*, pages 285–295, 2001. DOI: 10.1145/371920.372071 Cited on page(s) 41

[279] G. Sauer, H. Hochheiser, J. Feng, and J. Lazar. Towards a universally usable captcha. In *SOUPS*, 2008. Cited on page(s) 49

[280] T. Schulze. Worker motivation in crowdsourcing and human computation. In *HCOMP*, 2011. Cited on page(s) 61

[281] Scienceforcitizens.net. http://scienceforcitizens.net/. Cited on page(s) 52

[282] N. Seemakurty, J. Chu, L. von Ahn, and A. Tomasic. Word sense disambiguation via human computation. In *2nd KDD Human Computation Workshop*, 2010. DOI: 10.1145/1837885.1837905 Cited on page(s) 64

[283] Seti@home. http://setiathome.berkeley.edu/. Cited on page(s) 4

[284] B. Settles. Active learning literature survey. Technical report, University of Wisconsin-Madison, 2009. Cited on page(s) 6, 22, 23, 38

[285] B. Settles, M. Craven, and L. Friedland. Active learning with real annotation costs. In *NIPS Workshop on Cost Sensitive Learning*, 2008. Cited on page(s) 9

[286] H.S. Seung, M. Opper, and H. Sompolinsky. Query by committee. In *ACM Workshop on Computational Learning Theory*, pages 287–294, 1992. DOI: 10.1145/130385.130417 Cited on page(s) 23

[287] D. Shahaf and E. Amir. Towards a theory of ai completeness. In *8th International Symposium on Logical Formalizations of Commonsense Reasoning*, 2007. Cited on page(s) 21

[288] D. Shahaf and E. Horvitz. Generalized task markets for human and machine computation. In *AAAI*, 2010. Cited on page(s) 37, 74

[289] A. D. Shaw, J. J. Horton, and D. L. Chen. Designing incentives for inexpert human raters. In *CSCW*, 2011. DOI: 10.1145/1958824.1958865 Cited on page(s) 61

[290] V. Sheng, F. Provost, and P. Ipeirotis. Get another label? improving data quality and data mining using multiple, noisy labelers. In *KDD*, 2008. DOI: 10.1145/1401890.1401965 Cited on page(s) 23

[291] M. Shilman, D. Tan, and P. Simard. Cuetip: A mixed-initiative interface for correcting handwriting errors. In *CHI*, pages 323–332, 2006. DOI: 10.1145/1166253.1166304 Cited on page(s) 74

[292] M. Six Silberman, L. Irani, and J. Ross. Ethics and tactics of professional crowdwork. *XRDS*, 17(2):39 – 43, 2010. DOI: 10.1145/1869086.1869100 Cited on page(s) 74

[293] M. Six Silberman, J. Ross, L. Irani, and B. Tomlinson. Sellers' problems in human computation markets. In *HCOMP*, 2010. DOI: 10.1145/1837885.1837891 Cited on page(s) 74

[294] J. Silvertown. A new dawn for citizen science. *Cell*, 24(9), 2009. DOI: 10.1016/j.tree.2009.03.017 Cited on page(s) 52

[295] Y. Singer and M. Mittal. Pricing mechnaisms for online labor markets. In *HCOMP*, 2011. Cited on page(s) 22

[296] K. Siorpaes and M. Hepp. Games with a purpose for the semantic web. *IEEE Intelligent Systems*, 23(3):50–60, 2008. DOI: 10.1109/MIS.2008.45 Cited on page(s) 64

[297] D.B. Skillicorn and D. Talia. Models and languages for parallel computation. *ACM Computing Surveys*, 30(2):123–169, 1998. DOI: 10.1145/280277.280278 Cited on page(s) 17

[298] P. Smyth, U. Fayyad, M. Burl, P. Perona, and P. Baldi. Inferring ground truth from subjective labeling of venus images. In *NIPS*, pages 1–9, 1997. Cited on page(s) 26, 30

[299] R. Snow, B. O'Connor, D. Juraksky, and A. Ng. Cheap and fast – but is it good? evaluating non-expert annotations for natural language tasks. In *EMNLP*, 2008. DOI: 10.3115/1613715.1613751 Cited on page(s) 9, 35

[300] R. Speer, C. Havasi, and H. Surana. Using verbosity: Common sense data from games with a purpose. In *FLAIRS*, 2010. Cited on page(s) 65, 67

[301] K. Starbird. Digital volunteerism during disaster: Crowdsourcing information processing. In *CHI Workshop on Crowdsourcing and Human Computation*, 2011. Cited on page(s) 54

[302] N. Stewart, G.D.A. Brown, and N. Chater. Absolute identification by relative judgment. *Psychological Review*, 112:881–911, 2005. DOI: 10.1037/0033-295X.112.4.881 Cited on page(s) 59

[303] H. S. Stone. Algorithms, turing machines, and programs. In *Introduction to Computer Organization and Data Structures*. McGraw-Hill, 1972. Cited on page(s) 16

[304] A. Strehl and J. Ghosh. Cluster ensembles – a knowledge reuse framework for combining multiple partitions. *Journal of Machine Learning Research*, 3:583–617, 2002. DOI: 10.1162/153244303321897735 Cited on page(s) 31

[305] S. Sueken, J. Tang, and D. Parkes. Accounting mechanisms for distributed work systems. In *AAAI*, pages 860–866, 2010. Cited on page(s) 62

[306] B.L. Sullivan, C.L. Wood, M.J. Iliff, R. E. Bonney, D. Fink, and S. Kelling. ebird: A citizen-based bird observation network in the biological sciences. *Biological Conservation*, 2009. DOI: 10.1016/j.biocon.2009.05.006 Cited on page(s) 51, 52

[307] S. Suri and D. Watts. Cooperation and contagion in networked public goods experiments. *PLoS ONE*, 6(3), 2011. DOI: 10.1371/journal.pone.0016836 Cited on page(s) 46

[308] P. Sweetser and P. Wyeth. Gameflow: a model for evaluating player enjoyment in games. *ACM Computers in Entertainment*, 3(3), 2005. DOI: 10.1145/1077246.1077253 Cited on page(s) 63

[309] G. Takacs, I. Pilaszy, B. Nemeth, and D. Tikk. On the gravity recommendation system. In *KDD*, pages 22–30, 2007. Cited on page(s) 31

[310] J. Tam, J. Simsa, S. Hyde, and L. von Ahn. Breaking audio captchas. In *NIPS*, 2009. Cited on page(s) 49

[311] O. Tamuz, C. Liu, S. Belongie, O. Shamir, and A. Kalai. Adaptively Learning the Crowd Kernel. In *Proceedings of the 28th International Conference on Machine Learning*, 2011. Cited on page(s) 74

[312] L. Terry, V. Roitch, S. Tufail, K. Singh, O. Taraq, W. Luk, and P. Jamieson. Harnessing human computation cycles for the fpga placement problem. In *ERSA*, 2009. Cited on page(s) 10

[313] L. Terveen and D. W. McDonald. Social matching: a framework and research agenda. *TOCHI*, 12(3):401–434, 2005. DOI: 10.1145/1096737.1096740 Cited on page(s) 41

[314] S. Thaler, E. Simperl, K. Siorpaes, and C. Hofer. A survey on games for knowledge acquisition. Technical report, STI, May 2011. Cited on page(s) 69

[315] The great sunflower project. `http://www.greatsunflower.org/`. Cited on page(s) 52

[316] The lost tomb of genghis khan. `http://exploration.nationalgeographic.com/mongolia/`. Cited on page(s) 8

[317] Trec. `http://trec.nist.gov/`. Cited on page(s) 9

[318] D.J. Trumbull, R. Bonney, D. Bascom, and A. Cabral. Thinking scientifically during participation in a citizen-science project. *Science Education*, 84:265–275, 2000. DOI: 10.1002/(SICI)1098-237X(200003)84:2%3C265::AID-SCE7%3E3.0.CO;2-5 Cited on page(s) 54

[319] D.J. Trumbull, R. Bonney, and N. Grudens-Schuck. Developing materials to promote inquiry: Lessons learned. *Science Education*, 89:879–900, 2005. DOI: 10.1002/sce.20081 Cited on page(s) 54

[320] G. Tullock. Efficient rent seeking. In *Towards a Theory of the Rent-Seeking Society*. Texas A&M University Press, 1980. Cited on page(s) 62

[321] A. M. Turing. Computing machinery and intelligence. *Mind LIX*, 2236:43–60, October 1950. Cited on page(s) 3

[322] Turkit. `http://groups.csail.mit.edu/uid/turkit/`. Cited on page(s) 19, 21

[323] D. Turnbull, L. Barrington, D. Torres, and G. Lanckriet. Five approaches to collecting tags for music. In *ISMIR*, 2008. Cited on page(s) 8

[324] D. Turnbull, L. Barrington, D. Torres, and G. Lanckriet. Semantic annotation and retrieval of music and sound effects. *TASLP*, 16(2):467–476, 2008. DOI: 10.1109/TASL.2007.913750 Cited on page(s) 8

[325] D. Turnbull, R. Liu, L. Barrington, and G. Lanckriet. A game-based approach for collecting semantic annotations for music. In *ISMIR*, pages 535–538, 2007. Cited on page(s) 9, 51, 64

[326] A. Tversky and D. Kahneman. Judgment under uncertainty: Heuristics and biases. *Science*, 185:1124–1130, 1974. DOI: 10.1126/science.185.4157.1124 Cited on page(s) 59

[327] J. S. Uebersax. Statistical modeling of expert ratings on medical treatment appropriateness. *Journal of the American Statistical Association*, 88(422):421–427, 1993. DOI: 10.2307/2290320 Cited on page(s) 29

[328] D. Vickrey, A. Bronzan, W. Choi, A Kumar, J. Turner-Maier, A. Wang, and D. Koller. Online word games for semantic data collection. In *EMNLP*, 2008. DOI: 10.3115/1613715.1613781 Cited on page(s) 64

[329] D. Vickrey, A. Bronzan, W. Choi, A. Kumar, J. Turner-Maier, A. Wang, and D. Koller. Online word games for semantic data collection. In *EMNLP*, pages 535–538, 2008. Cited on page(s) 67, 69

[330] S. Vijayanarasimhan and K. Grauman. What's it going to cost you?: Predicting effort vs. informativeness for multi-label image annotations. *CVPR*, pages 2262–2269, 2009. DOI: 10.1109/CVPR.2009.5206705 Cited on page(s) 23

[331] A. Vlachos. Active annotation. In *EACL Workshop on Adaptive Text Extraction*, 2006. Cited on page(s) 22

[332] L. von Ahn. *Human Computation*. PhD thesis, Carnegie Mellon University, 2005. Cited on page(s) 3

[333] L. von Ahn, M. Blum, and J. Langford. Telling humans and computers apart automatically: How lazy cryptographers do ai. *Communications of the ACM*, 47(2):57–60, 2004. Cited on page(s) 48, 49

[334] L. von Ahn and L. Dabbish. Labeling images with a computer game. In *CHI*, pages 319–326, 2004. DOI: 10.1145/985692.985733 Cited on page(s) 8, 25, 49, 63, 67, 69

[335] L. von Ahn and L. Dabbish. Designing games with a purpose. *Communications of the ACM*, 51(8), August 2008. DOI: 10.1145/1378704.1378719 Cited on page(s) 65, 67

[336] L. von Ahn and L. Dabbish. General techniques for designing games with a purpose. In *CACM*, pages 58–67, 2008. Cited on page(s) 7, 63

[337] L. von Ahn, R. Liu, and M. Blum. Peekaboom: A game for locating objects in images. In *CHI Notes*, pages 55–64, 2006. DOI: 10.1145/1124772.1124782 Cited on page(s) 65, 69

[338] L. von Ahn, R. Liu, and M. Blum. Verbosity: A game for collecting common-sense knowledge. In *CHI Notes*, pages 75–78, 2006. DOI: 10.1145/1124772.1124784 Cited on page(s) 25, 65, 69

[339] L. von Ahn, B. Maurer, C. McMillen, D. Abraham, and M. Blum. recaptcha: Human-based character recognition via web security measures. *Science*, pages 1465–1468, 2008. DOI: 10.1126/science.1160379 Cited on page(s) 49

[340] J. von Neumann. Probabilistic logics and the synthesis of reliable organisms from unreliable components. In C. E. Shannon and J. McCarthy, editors, *Automata Studies*, pages 43 – 98. Princeton University Press, 1956. Cited on page(s) 21

[341] B.C. Wallace, K. Small, C.E. Brodley, and T.A. Trikalinos. Who should label what? Instance allocation in multiple expert active learning. In *Proc. of SIAM International Conference on Data Mining (SDM)*, 2011. Cited on page(s) 23

[342] J. Wang, S. Faridani, and P. Ipeirotis. Estimating the completion time of crowdsourced tasks using survival analysis models. In *CSDM*, 2011. Cited on page(s) 40

[343] I. Weber, S. Robertson, and M. Vojnovic. Rethinking the esp game. In *CHI*, pages 3937–3942, 2009. DOI: 10.1145/1520340.1520597 Cited on page(s) 67

[344] G.M. Weiss and F. Provost. Learning when training data are costly: the effect of class distribution on tree induction. *Journal of Artificial Intelligence Research*, 19:315–354, 2003. DOI: 10.1613/jair.1199 Cited on page(s) 23

[345] P. Welinder, S. Branson, S. Belongie, and P. Perona. The multidimensional wisdom of crowds. In *NIPS*, pages 1–9, 2010. Cited on page(s) 28, 29

[346] P. Welinder and P. Perona. Online crowdsourcing: rating annotators and obtaining cost-effective labels. In *CPVR*, 2010. Cited on page(s) 29, 40

[347] S.C. Weller. Cross-cultural concept of illness: Variation and validation. *American Anthropologist*, 86(2):341–351, 1984. DOI: 10.1525/aa.1984.86.2.02a00090 Cited on page(s) 27

[348] S.C. Weller. Cultural consensus theory: Applications and frequently asked questions. *Field Methods*, 19(4):339–368, 2007. DOI: 10.1177/1525822X07303502 Cited on page(s) 29

[349] S.C. Weller and N.C. Mann. Assessing rater performance without "gold standard" using consensus theory. *Medical Decision Making*, 17(1):71–79, 1997. DOI: 10.1177/0272989X9701700108 Cited on page(s) 27

[350] J. Whitehill, P. Ruvolo, T. Wu, J. Bergsma, and J. Movellan. Whose vote should count more: Optimal integration of labels from labelers of unknown expertise. In *NIPS*, 2009. Cited on page(s) 29

[351] Wikiepdia entry for "list of cognitive biases". http://en.wikipedia.org/wiki/List_of_cognitive_biases. Cited on page(s) 59

[352] Wikipedia. http://wikipedia.com. Cited on page(s) 5

[353] T.R. Williams. Reconsidering the history of the aavso. *Journal of the American Association of Variable Star Observers*, 29:132, 2001. Cited on page(s) 51

[354] R. Wilson and F. Keil, editors. *The MIT Encyclopedia of the Cogntivie Sciences*. MIT Press, 1999. Cited on page(s) 29

[355] J. Wolfers and E. Zitzewitz. Prediction markets. *Journal of Economic Perspectives*, 18(2):107–126, 2004. DOI: 10.1257/0895330041371321 Cited on page(s) 32

[356] J. Yan and A.S. El Ahmad. Usability of captchas. In *SOUPS*, 2008. DOI: 10.1145/1408664.1408671 Cited on page(s) 49

[357] Y. Yan, R. Rosales, G. Fung, and J. Dy. Active learning from crowds. In *ICML*, 2011. Cited on page(s) 29

[358] A. Yates and O. Etzioni. Unsupervised methods for determining object and relation. *Journal of Artificial Intelligence Research*, 34:255–296, 2009. Cited on page(s) 30

[359] Y. Yue and T. Joachims. Interactively optimizing information retrieval systems as a dueling bandits problem. In *ECML*, 2009. DOI: 10.1145/1553374.1553527 Cited on page(s) 74

[360] O. F. Zaidan and C. Callison-Burch. Feasibility of human-in-the-loop minimum error rate training. In *EMNLP*, 2009. DOI: 10.3115/1699510.1699518 Cited on page(s) 32

[361] H. Zhang, E. Horvitz, Y. Chen, and D. Parkes. Task routing for prediction tasks. In *ACM EC Workshop on Social Computing and User Generated Content*, 2011. Cited on page(s) 41

[362] H. Zhang, E. Horvitz, R. Miller, and D. Parkes. Crowdsourcing general computation. Technical report, Microsoft Research, 2011. Cited on page(s) 19

[363] J. Zhang, M. S. Ackerman, and L. Adamic. Expertise networks in online communities: Structure and algorithms. In *WWW*, 2007. DOI: 10.1145/1242572.1242603 Cited on page(s) 41

[364] Y. Zhou, G. Cong, B. Cui, C. Jensen, and J. Yao. Routing questions to the right users in online communities. In *ICDE*, 2009. Cited on page(s) 41

[365] Zooniverse. http://www.zooniverse.org. Cited on page(s) 51

Authors' Biographies

EDITH LAW

Edith Law is a Ph.D. candidate at Carnegie Mellon University, who is doing research on human computation systems that harness the joint efforts of machines and humans, with a focus on games with a purpose and citizen science. She is the co-organizers of the 1st and 3rd human computation workshops (HCOMP 2009 and HCOMP 2011), and the recipient of Microsoft Graduate Research Fellowship 2009-2011. Her work on TagATune has received a best paper nomination at CHI 2009.

LUIS VON AHN

Luis von Ahn is the A. Nico Habermann Associate Professor in the Computer Science Department at Carnegie Mellon University. His current research interests include building systems that combine the intelligence of humans and computers to solve large-scale problems that neither can solve alone. An example of his work is reCAPTCHA, in which over 750 million people—more than 10% of humanity—have helped digitize books and newspapers. He is the recipient of a MacArthur Fellowship, a Packard Fellowship, a Microsoft New Faculty Fellowship, and a Sloan Research Fellowship. He has been named one of the 50 Best Minds in Science by Discover Magazine, one of the 100 Most Creative People in Business of 2010 by Fast Company Magazine, and one of the "Brilliant 10" scientists of 2006 by Popular Science Magazine.